To my God-given wife and children,
To the Project Management fraternity across Africa
and the Diaspora,
I humbly dedicate this work.

Ignorance has an elder sister; her name is arrogance.
Once they hold hands, no book can put them asunder.

Managing Engineering Processes in Large Infrastructure Projects

Highly recommended reference material for all project practitioners such as project managers, engineering managers, project engineers, project controls managers, safety personnel, estimators, planners, and any person involved in major projects at executive level, including project directors and project sponsors. Given his strong engineering background, his excellent linguistic acumen, and decades of experience in the project environment, the author is very competent in sharing practical insights on the management of large infrastructure projects.
—Ilunga-Claude Tshimbidi (PhD), PrEng.
Former Chief Engineer at Sasol Technology.

Building and infrastructure projects are as important today as they have ever been. As the author counsels, it is important to respect the phases. The book is a stimulating yet technical guide on all the engineering phases of a project, commencing with a lucid exposition of the criticality of design. I recommend the book for students of engineering and practitioners alike, as well as those involved in the legal and financial aspects of infrastructure projects.
—C. King (Bert) Chanetsa, (LLB, Juris Doctor, FFin)
Partner Chanetsa Inc.; Author; Global Consultant; Researcher and Capacity Builder in Legal and Financial Aspects of Projects and Infrastructure Finance.

This book is a good guide to learning the fundamentals of project management from design to procurement and construction. I hereby recommend it to all the construction industry stakeholders such as architects, engineers, quantity surveyors, project managers, etc.
—Agoro Musiliu Olalekan (PhD Civil Engineering)
Chief Lecturer, Yaba College of Technology, Lagos.

Managing Engineering Processes in Large Infrastructure Projects

By

Pascal Bohulu Mabelo

Cambridge
Scholars
Publishing

Managing Engineering Processes in Large Infrastructure Projects

By Pascal Bohulu Mabelo

This book first published 2021

Cambridge Scholars Publishing

Lady Stephenson Library, Newcastle upon Tyne, NE6 2PA, UK

British Library Cataloguing in Publication Data
A catalogue record for this book is available from the British Library

ISBN (10): 1-5275-7504-7
ISBN (13): 978-1-5275-7504-2

TABLE OF CONTENTS

LIST OF FIGURES

LIST OF TABLES

FOREWORD

Over more than three decades of involvement in teaching and research in project management at tertiary level, I have become very much aware of the tremendous growth in not only the interest, but also the specific involvement of most business disciplines in project management as a clear and well-defined profession. This tremendous growth in awareness and interest has encouraged authors from across the globe to produce a library of textbooks covering every angle from the bare basics to the super-advanced side of project management.

A wide variety of literary support is thus available for new, aspiring and experienced project managers. However, having said this, the time is now right for more textbooks which will drill deeper into specific areas where our project management tribulations and experiences over the past few years have indicated some deficiencies (e.g., level of thinking, attitudes) and also areas where our project management maturity could possibly be improved.

What Pascal has achieved with this book is to create a novel, and may I say, an important angle of approach every project manager involved in large infrastructure projects should not only take note of, but should also become very competent in. Unfortunately, even in today's environment of well-developed project management principles and practices, we still find many large projects failing and/or missing their budgets, schedules and quality specifications. Examples such as the two power station megaprojects currently under construction in South Africa, where serious engineering mistakes have been identified, come to mind and highlight this problem very clearly.

Focussing on large engineering infrastructure projects, Pascal takes the project life cycle back to the initial design phase; he points out that errors made in this phase of the project are often the root cause of many minor mishaps and even major failures. He links the concepts of systems engineering, operational requirements and project management together into a process of "managing engineering" with the goal of closely interconnecting the client, the designers, the contractors and the operators into a collaborative team focussing on the ultimate success of the project from initial design to final handover.

With this book, Pascal takes both his academic-orientated as well his business-orientated readers on a journey of the theory and practices of project managing large engineering infrastructure projects. This journey spans the engineering design and development of large and complex infrastructure projects right from the conceptual design phase to the detailed design and specification phase and follows on to the construction and implementation phases, each with its own specific life cycle requiring planning, monitoring and control — but still fully aligned with the total cradle-to-grave life cycle, or "*from womb to tomb*" as the author aptly calls it.

With Pascal's 25+ years of practical experience in the engineering industry, where he was very much involved in not only the technical design and reviews of major infrastructure projects but also the project and facilities management of such projects, he has gained the experience and expertise to now share his knowledge with us in this book, just as he has done in previous publications.

Deon Kruger

Deon Kruger is a Senior Lecturer at the University of Johannesburg (South Africa), Faculty of Engineering and the Built Environment. With 30+ years of experience as a researcher and professional engineer with a strong passion for project management, Deon Kruger was elected Fellow of the South African Institute of Civil Engineers, then Vice-President of the influential International Congress on Polymers in Concrete. He sure has a great appreciation of engineering.

PREFACE

Engineering is generally defined as the design, construction and operation of efficient and economical structures, equipment and systems. Thus, Engineering Design is an essential element of the system life cycle. No matter how good the manufacturing, production, sales and marketing might be, if a product/system is poorly designed, the end product will ultimately fail. Further, engineering mistakes often result in massive overruns and/or poor operability. Indeed, no project team can control time and/or costs once they fail to manage the engineering endeavour.

Moreover, engineers are humans and, therefore, fallible in their endeavours. In fact, no engineering, being man-made, should assume qualities beyond that of humans, for not everything we have created has been successful. It is misplaced to expect that engineers will never make mistakes in their duties. The poor performance of individuals should not be the issue; rather, what is required is a set of procedures, processes and organisational arrangements that will prevent costly engineering errors from occurring in the first place.

The US Department of Defence (DoD) have published the Military Standard: Engineering Management [MIL-STD-499A (USAF)] to assist Government and contractor personnel in defining the systems engineering effort in support of defence acquisition programmes. The fundamental concept of this Standard was to present a single set of criteria against which all may propose their individual internal procedures as a means of satisfying engineering requirements. Yet, even such a seminal work would need to be expanded (to better accommodate large/complex infrastructure projects) and be translated into a *step-by-step* elaboration that further provides the reader with its Systems Engineering rationale as well as any pertinent methods that support the practical applications of such a Standard.

This book basically introduces and discusses engineering approaches that can guide engineers in their design and development efforts, particularly when it comes to large and complex infrastructure projects. Humans created all shapes and sizes of "drums" way before they could understand the physics of sound (therefore, one concedes that "*To engineer is human*"). However, the design and development of today's infrastructure systems such as rail and road networks, power generation plants, water treatment plants, and complex shopping malls would not be achieved except through processes and methods arising from Systems Thinking or Systems Engineering concepts and principles. Relationships with other systems and the environment should be considered.

> "Unless we look at things as systems that are somewhat interconnected, we wouldn't perceive anything beyond their faint and fleeting shadows" (The author).

Furthermore, systems are generally "*created for humans*". This indeed applies to large and complex infrastructures that not only support the provision of essential services (e.g., water, energy, transport), but should also satisfy other "emotional" needs, including the appreciation of beauty and *amour-propre*. From that perspective, Engineering Design and Development ought to rather adopt human-centric approaches that call for the blending of art and science. As succinctly discussed in this book, Design Thinking, the systems-way of "Thinking about Design", caters for such.

ABOUT THE AUTHOR

Pascal Bohulu Mabelo *MBA, MSc (Eng), Pr Eng (South Africa), Pr CPM (South Africa), PMP.*

Pascal possesses a wide range of technical and managerial skills pertaining to infrastructure projects and has had the opportunity to work on an array of large-scale projects with people from various backgrounds. He presently works as a principal consultant through his own practice, E 6 Project Consulting (Pty) Ltd, and contracts on diverse assignments in the private and public sectors. Pascal is also the author of "How to Manage Project Stakeholders – *Effective Strategies for Large Infrastructure Projects*" (2020) and "Operational Readiness – *How to Achieve Successful System Deployment*" (2020). Both books are published by Routledge.

Pascal has worked in large projects as a design engineer, project/programme manager, project consultant, and Project Management executive. He established and, for more than eight years, headed up the Project Management Centre of Excellence (PM-CoE) at the capital-projects division of a multi-billion Rand company in South Africa. He has chaired hundreds of gateway reviews of infrastructure projects, from medium-sized to large and complex. With more than 25 years of professional experience in the industry, he has steadily built up an extensive network and was honoured to serve as the national chairman of Project Management South Africa (PMSA), the leading Project Management professional association in Southern Africa.

Pascal is also well known in the industry for having authored several articles in Project Management magazines and journals, and has guest lectured on project-related processes at various top-ranked South African universities, not to mention lecturing Operations Management at an international business school as well as speaking at project management conferences, seminars, and webinars. He is currently promoting the application of Systems Thinking and/or Systems Engineering principles and concepts to unravel complexity in Large Infrastructure Projects (LIPs) in order to address their persistent risks of failure and their massive, and even pernicious, cost and schedule overruns.

ABOUT THIS BOOK

Having read the book "*To Engineer is Human*", one not only gets the sense that "engineering" is a human endeavour, but that it is also actually inherent to human nature; hence, we humans are engineers and designers by nature. Ordinary people have throughout their lives learned to engineer and design things in some way—from the pre-school boy building a sandcastle on the beach, to the woman in rural Africa folding an old garment into the "circular pad" she puts on her head to even up and cushion the "loading" of the firewood she carries. We ought to only call in the professionals for more complicated tasks. Thus, Engineering Design is not as esoteric as many might make it out to be. Sir James Dyson was not a *trained* engineer; yet, he invented the "cyclonic" vacuum cleaner and many other useful devices.

The humanness of Engineering Design, nevertheless, will incidentally entail that most engineers and designers are fallible and expected to make errors, which at times may prove very detrimental, if not catastrophic, to the system being developed. In addition, the Engineering Design and Development (EDD) process is traditionally approached in a rather implicit and almost discretionary manner which, as a result, often makes its management more difficult than it ought to be.

Considering that engineers are humans and, akin to many other professions, the consequences of their errors will most likely include adverse, disastrous implications to society, there is a necessity to provide engineers with an arsenal of principles, techniques, practices and processes that will help them to manage the whole Design and Development endeavour in projects.

Engineering Design and Development is important to those organisations that are involved with infrastructure design, either as its consumers or producers. As consumers, engineering managers are part of the owner's team that purchases the large and complex infrastructure system (to be designed by others); however, as producers, they are part of the design team. Either way, those engineering managers will be involved in managing the design of product or system design; either way, their mastery of engineering will be tested—therefore, those engineers shall benefit from any *guidance* on managing EDD processes.

Engineering Design and Development being a vital component of project delivery and Asset Management processes, it becomes critical that this particular set of activities is planned at the same level of definition and, thus, subsequently executed with the same diligence as any other technical life cycle activities. This means engineering activities ought to be executed in accordance with a plan detailing at least five key delivery aspects such as:

> (1) Scope of work in terms of engineering activities/deliverables; (2) Timelines and sequencing; (3) Organisational structure, teams and skills; (4) Resources required, e.g., financial, personnel, equipment; and (5) Major risks identified at that stage.

Moreover, as the complexity of the product/system increases, a greater priority and attention to detail needs to be given to the construction and/or fabrication process (i.e., "*How to build*").

In many infrastructure projects (e.g., long bridge over an extreme height), the design of the "*How to build*" is as important and as complex as the design of the "*What to build*". One may think of the complex planning required to build the Duge Beipanjiang (China), due to the vertical distance between the bridge and the bottom of the valley it traverses being equal to the height of a 200-storey skyscraper—and took 39 months to complete.

In another, somewhat infamous, instance the architects of the iconic Sydney Opera House were allowed total freedom in their novel designs (i.e., "*What to build*"). However, from the time the government insisted on construction getting underway, there soon appeared to be problems due to the construction methodology/process (i.e., "*How to build*") still being developed. These Engineering Management inadequacies proved problematic, if not controversial. Thus, plans and designs had to be modified during construction—resulting in 10 years of delays and an outlandish 1457 % cost overrun (viz., from an initial budget of AUS $ 7 million to a Total Spend of AUS $ 102 million at its completion in 1973).

Other equally insidious Engineering Management inadequacies include:

- Relying more on the engineers' abilities than on effective processes
- Solving the wrong problem (i.e., misdiagnosis, tackling symptoms)
- Using non-systemic methods to design Complex Adaptive Systems
- Commencing Construction before Engineering Design is complete
- Failing to align Engineering with other project life cycle processes
- Failing to plan Engineering activities (i.e., scope, schedule, budget) at the same level of definition as any other project delivery activity.

This book is therefore intended to not only address such inadequacies, but also to concomitantly achieve two very important capacitation objectives pertaining to the effective delivery of any megaprojects, namely:

(i) The capability of the engineering (i.e., EDD) team working on Large Infrastructure Projects in devising and/or adhering to a structured process when designing the System-of-Interest; hence, making the process explicit and consistent. This would permit an effective management not only of the design and development tasks, but of their integration into other project life cycle activities, including adherence to scope, timeline and budget provisions. No engineering manager shall again wonder what an Engineering Plan is. (A "Head of Project Engineering" once asked the author in a project meeting, in front of his team, "*Tell me, what is an Engineering Plan?*")

(ii) The capacity of "non-engineer" members of the project team to understand the purpose and requirements of the broader engineering (EDD) process in order to meaningfully *support and/or participate* in the design and development activities, as well as *accommodate* such activities and deliverables in their own scope/tasks.

It is hoped that meeting the above objectives will vastly benefit the LIPs industry. The aforesaid capability and capacity will allow not only the project manager and the engineering manager, but also the Chief Executive Officer (CEO) and the Chief Projects Officer (CPO) to "*think alike*" (just as great minds should) and "*speak the same language*" when it comes to capital projects (e.g., large and complex infrastructure projects) and other similar initiatives.

This book might be the closest to an "*Engineering for Non-engineers*" tome. While non-engineering readers are expected to grasp at least half of the content of this material, those sufficiently versed in engineering practices (who might already be proficient in what is discussed herein) would still benefit from the *advanced* half, particularly with regards to the processes, tasks and resources required to attain a successful design, and how to manage Engineering Design as a technical endeavour in any large and complex projects.

ACRONYMS

01	BC	Before Christ
02	CAS	Complex Adaptive System
03	CASoS	Complex Adaptive System of Systems
04	CO_2	Carbon Dioxide
05	ConOps	Concept of Operations
06	COTS	Commercial Off The Shelf
07	DC	Direct Current
08	DoD	Department of Defence (USA)
09	EDD	Engineering Design and Development
10	FEL	Front End Loading
11	GPS	Global Positioning System
12	HVAC	Heating Ventilation and Air Conditioning (System)
13	ICT	Information and Communications Technology
14	INCOSE	International Council on Systems Engineering
15	ISO	International Organization for Standardization
16	ITS	Intelligent Transportation System
17	LIP	Large Infrastructure Project
18	MIE	[Dutch] Ministry of Infrastructure and the Environment
19	NASA	National Aeronautics Space Agency
20	NPV	Net Present Value
21	PMBoK	Project Management Body of Knowledge
22	Q_lRA	Qualitative Risk Assessment
23	Q_nRA	Quantitative Risk Assessment
24	SE	Systems Engineering
25	SE-PLM	Systems Engineering-based Project Life Cycle Methodology
26	SoI	System of Interest
27	SoS	System of Systems

CHAPTER 1

ENGINEERING AND SYSTEMS THINKING

The art and science of "engineering" has at its heart not the existing world, but the world that engineers themselves create. For while honeybees have been making the same kind of hive for centuries, human inventions and structures are in a state of constant and rapid evolution. Humans like change—at times, just for the sake of it. Their tastes, ambitions, and resources are ever changing; they like bigger, taller or longer things in ways that honeybees do not or cannot.

Petroski laments in the same vein, "It is as if engineers and non-engineers alike, being human, want their creations [i.e., products of designs] to be superhuman", forgetting that, "Because man is fallible, so are his constructions, however". No manner of engineering, being man-made, can assume qualities beyond those of humans. Nevertheless, "All of these extra-engineering considerations [usually] make the task of the engineer perhaps more exciting and certainly less routine than that of [honeybees or ants]" (Petroski, 1992).

Engineering Design is defined as, "The process of devising a system, component, or process to meet [the] desired needs (of some customers). It is often an iterative [i.e., recursive and fractal] decision-making process, in which the basic sciences, mathematics, [arts,] and engineering sciences are applied to optimally convert resources to meet a stated objective" (Haik and Shahin, 2011)—sections of this book discuss artistic *apports* to engineering.

Among the fundamental elements of the design process are the setting up and establishment of objectives and criteria, along with synthesis, analysis, construction or fabrication, testing and evaluation (Haik and Shahin, 2011).

Since Engineering Design is applied "*to optimally convert 'resources' to meet a stated objective*", it ought to address as one of its fundamental elements the "management" of such resources (e.g., financial, skills, capacity, equipment, time, etc.). Herein lies one of the difficulties with most engineering design methodologies—they do not allow for such management.

Engineering Design is an essential element of the system life cycle. To be sure, "No matter how good the manufacturing, production, sales, etc. are, if a product is poorly designed, the end product will still be a bad idea and will ultimately fail, as no one likes to purchase a bad idea. Not all that engineers build becomes successful; occasionally, catastrophic failures occur […]" (Haik and Shahin, 2011). In LIPs, such failures arise from design errors about the "What to build" (i.e., system and its performance) as well as the "How to build" (i.e., building process) during the FEL (Front End Loading) phases of Conceptual (FEL-1), Pre-Feasibility (FEL-2), and Feasibility (FEL-3). They only become manifest in Construction or Operations, though.

"To Engineer is Human", Petroski argues. Engineering is definitely a human endeavour and, thus, is prone to errors that may cause one to ask, "Where is our progress?" (Petroski, 1992). "Decision-making in projects is found to lack in definition […] is often made by engineering judgements and […] by perpetuating decisions made on other similar projects […] [thus] bad decisions follow project teams" (Krauss, 2014). Shall one celebrate success despite so many "technological embarrassments"?

The prevalence of engineering design errors and the associated cumulative negative impact upon the structural integrity of buildings and the financial performance of projects and organisations is a recurrent predicament within the construction and engineering industry.

> "Design errors are a major cause of accidents and research has revealed that gross errors can cause 80 to 90% of the failures occurring on buildings, bridges and other civil engineering structures. Despite the considerable amount of research, [...] design errors remain a constant threat [particularly to the large and complex infrastructure sector]." (Love et al, 2011)

The pernicious link between design errors, the integrity of the structure, and the financial performance of the organisation is not always obvious. This nexus only becomes obvious in the event of failure. Hence, Haik and Shahin (2011) have highlighted two well-publicised "disasters" associated with engineering systems as follows:

(i) A skywalk at the Kansas City Hyatt Regency Hotel collapsed just after the hotel opened in 1981. The designer had failed to consider the *dead-weight* of the structure. The skywalk rods were not designed to hold the combined weights of the walkways and the 2000 people that had gathered on them; 200 people were injured, and 114 were killed.

(ii) The Chernobyl nuclear power plant disaster of 1996 was also the result of a design error (i.e., flawed reactors using graphite to sustain the chain reaction, but water instead of an inert gas to cool the nuclear fuel). The calamity led to the evacuation and resettlement of over 336,000 people, caused 56 direct deaths, and more than 4000 thyroid cancer cases among children. Approximately 6.6 million people were highly exposed to radiation.

Worse yet, the Fukushima nuclear disaster arose from incorrect assumptions that the engineering of the facility would exhibit no *vulnerabilities* to tsunamis. These disasters caused the loss of lives, and financial and reputational damages. Measures should have prevented such *insidious* errors from arising "*in the first place*", but they clearly didn't!

"*Errare humanum est, sed perseverare diabolicum*" [Latin]; to err is human, but to persevere (proceed in error) is diabolic or fatal. This maxim applies to engineering too. Furthermore, according to Haik and Shahin (2011), the leading causes of failure in most engineering endeavours are as follows:

- Faulty reasoning from good design assumptions
- Incorrect and/or over-extended design assumptions
- Poor understanding of the problem to be solved
- Errors in design calculations and/or in drawings
- Incorrect and/or inconsistent design specifications
- Faulty manufacturing and/or assembly
- Incomplete experimentation and inadequate data collection

An effective Engineering Design and Development process must, therefore, cater for "independent third party" checks and reviews at crucial stages during its implementation.

> "[Most] Design errors can be significantly reduced when design checks are undertaken prior to construction commencing [...] through rigorous [better still, systemic] design checks, 32% of errors can be detected. In addition, if an independent third party is used then as much as 55% of design errors could be accounted for [...] Humans are fallible, errors must be expected and individuals' poor performance is a non-issue. Instead, the focus [of error prevention] should be on the failure in procedures, processes, teams and the organization." (Love et al, 2011)

Furthermore, although infrastructure comes second to business operations for many organisations, the delivered project has a significant operational component. Projects are primarily for the sake of "operations", in order to improve (or establish) the operational environment. Hence, the operational requirements (e.g., safety, volume and timing of ramp-up) are paramount and need to be introduced into the design from project inception and for each design review thereafter (Australian Government, 2012). Anything else (i.e., assets failing to improve operations) will prove deleterious.

However, in most "traditional" Design Processes, decisions concerning the technical solutions to problems are often made by the designers themselves, who implicitly select preferred options on the basis of their own understanding. For example, the engineer will indulge in some "*design-by-force-of-habit*" paradigm.

Even among professional design engineers, there is a tendency to revert to the latest design (to repeat a success?) rather than undertake to design the system or structure from scratch. To make things worse, swinging the pendulum to the other extreme, many such design engineers will get caught up in seeking to create their "masterpiece", without any consideration of the owners' requirements nor of the resource constraints.

It may seem that many design engineers are driven by their own ambitions or frustrations, and the belief that, "*The market does not know what they want; it is the engineer's job to tell them*". Those designers will proceed headlong in their design endeavours *no matter what*, until they triumph.

Their engineering process is neither *explicit* nor *consistent* (thus, making its management difficult) and would not offer a structural Verification and Validation (V&V)—this situation is problematic. It is crucial to establish the adequacy of the design (whether the design suits and solves the problem) and the efficiency of the design solutions even as follows:

(1) Does the solution meet the "requirements" to the maximum possible extent?
(2) Does the solution afford the best possible "quality–price" relationship?
(3) Is there a distinct "structural separation" (i.e., clear at any point in time) between the requirements and the solutions in the life cycle methodology?
(4) What are the "Tasks and/or deliverables, resources and key risks" involved in converting the (set of) requirements into solutions (to the problem)?

Not many design engineers will be able to answer these important questions. Furthermore, "In the past, development engineers did not work as a system. If one team solved a problem, it usually created a problem for another team.

For example, a development team was trying to reduce the noise and vibration for a new Lincoln Continental. They solved their problem by adding weight to the braking system, thereby creating weight and structure problems for the braking system team" (Bellingham, 2001). This attests to how design engineers are prone to falling into 'fixes that fail' and 'shifting of the burden' architypes or traps:

> Fixes that fail—i.e., investment towards solving a 'perceived' (or 'symptoms' of a) problem in a particular entity, but since the underlying causes were not understood, the effort (i.e., solution) might *unintendedly* lead to more problems in another entity, even creating some dependence on the fix (i.e., cure) thus developed (Meadows et al, 1972; Meadows et al, 2009).
>
> Shifting of the burden—i.e., performance improvement in a particular entity might cause the deterioration of performance in two or more other entities in the same system, but not necessarily in proximity of time and location (Meadows et al, 1972; Meadows et al, 2009).

In light of the foregoing discussions and considerations, design engineers will benefit from a design method that is not only "structured" (i.e., a replicable process), but that also provides a means of detecting and preventing errors. A process is required that can transform a problem into a solution, or a dream into reality; this is about a process that determines what should be done, who could get involved, how much it might cost, and how long it would take, and most importantly, whether the envisaged system will be successfully delivered. Such a process would be technical in its essence, at least in its final outcomes. Thus, this book on *Managing Engineering Processes* is intended to fill this gap.

Indeed, Slocum notes that:

> "Long before any design project starts, the design engineer has to believe that there is a problem that is worthy of their attention. The design engineer must *feel* a need to solve the problem [i.e., empathy] [...] must have a *yearning* to solve the problem [i.e., an undesirable situation] [...] However, one must be very careful about managing one's passion, lest one's *excitedness* overshadows true opportunity. In the world of business, it does not matter if the design engineer passionately creates a product that does not meet customer needs. Passion means little if the design is tainted by ignorance and inattention to detail." (Slocum, 2008)

It follows that the required Engineering Design and Development (EDD) process is, "A series of steps [design] engineers use to guide them in problem-solving. Engineers must ask a question, imagine a solution, plan a design, create that model, experiment and test that model, then take time to improve the original [solution]" (NASA, 2011).

Furthermore, the government of Alberta (Australia) proposes an "*Integrated Design Process* [that] is a collaborative team approach with the client group, including occupants and operating staff, and a multidisciplinary design team, focusing on the design, construction, operation, and occupancy of a building over its complete life cycle" (Standards and Specifications Specialist, 2018).

They strongly encourage this Integrated Design Process, "For all projects [i.e., including Large Infrastructure Projects], as the functional, environmental, and economic goals are better defined and realized by proceeding from whole building system strategies, through increasing levels of specificity, to achieve more optimally integrated solutions. The integrated design process is more effective the earlier it begins in the project timeline." (Standards and Specifications Specialist, 2018).

The author has tackled this vital aspect of project delivery in the following prior publications, "*How to Manage Project Stakeholders*" (Mabelo, 2020) and "*Operational Readiness*" (Mabelo, 2020b). In fact, engaging stakeholders during the design development phase assists in incorporating functional integration (of subsystems) into the design phase and in ensuring minimum disruption to the community during construction (Australian Government, 2012).

Hence, by considering the requirements of the "systems of interest" (SoI) over the life cycle (i.e., *from womb to tomb*) during the Engineering Design process, the "Holistic System Designer" is able to focus the solution(s) on producing maximum performance and quality (and/or efficiency) for the entire life cycle. The concept of multiple solutions is one of the differences between the "Holistic Thinking" Systems Engineering (SE) approach to problem-solving and the "traditional" problem-solving that identifies one solution and then runs with it (Kasser, 2015). Moreover, SE makes it possible to create transparency in terms of the impact of making changes to requirements and preconditions.

Bearing in mind the foregoing deliberations, it is not surprising that SE is defined as:

> "An interdisciplinary approach and means to enable the realization of successful systems. It focuses on defining customer needs and required functionality early in the development cycle, documenting [every set of]

> requirements, and then proceeding with design synthesis and system validation while considering [i.e., perceiving] the complete problem: operations, cost and schedule, performance, training and support, test, manufacturing, [upgrades,] and disposal. SE considers both the business and the technical needs of all customers with the goal of providing a quality product that meets users' needs." (Fossnes and Forsberg, 2006)

Therefore, the point is ultimately about satisfying operational requirements effectively. SE, indeed, provides an integrated and structured set of methodologies for successfully implementing and managing projects. Thus, according to Dutch MIE (ProRail, 2013), the core elements (including Engineering Design) extracted from the definition used to describe SE could be listed as follows:

(i) Structured specification of requirements
(ii) Structured design of suitable solution to requirements
(iii) Use of proper approach to produce this solution
(iv) Use of proper approach to manage the produced solution (or system)
(v) Use of the controlled "approach to manage the total system" during its entire life cycle, including proper V&V

While most of the above aspects will have a deep bearing on the engineering process, it is particularly items (ii) and (iii) that constitute the essence of Engineering Design and Development (EDD). It is not surprising that "*a suitable solution to requirements*" and "*a proper approach to produce this solution*" both form the thrust of this book.

In the same vein, Ryschkewitsch et al (2009) argue:

> "Systems engineering is holistic and integrative. It incorporates and balances [all] the contributions of structural, mechanical, electrical, software, systems safety, and power engineers, plus many others, to produce a coherent whole [...] Systems engineering is not only about the details of requirements and interfaces among subsystems [...] Systems engineering is first and foremost about getting the right design – and then about maintaining and enhancing its technical integrity, as well as managing complexity with good processes to get it right".

Systems Engineering (SE) ought to direct System Design, which makes relevant the concept of "Systems Thinking" Design. "However," Kasser (2007) cautions, "poor early stage systems engineering does seem to have been a contributor to some of those failures resulting from producing solutions systems that do not remedy the need when deployed".

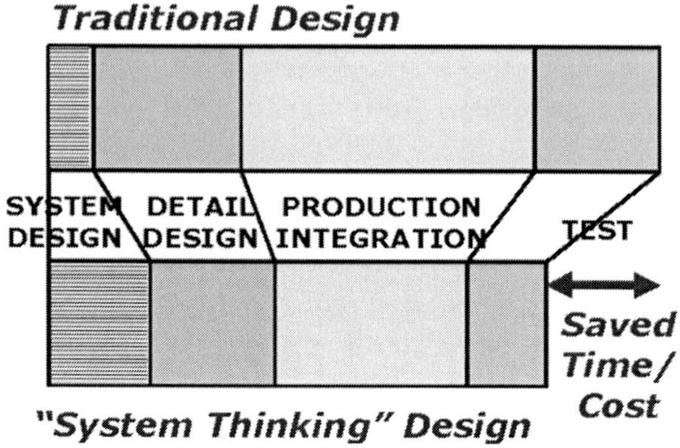

Figure 1.1 – Cost Savings by "Systems Thinking" Design (Adapted: Honour, 2004)

SE is first and foremost about "*getting the right design*", as Ryschkewitsch so eloquently puts it. And Honour (2004) argues that a greater emphasis on the System Design creates easier, more rapid integration and testing—the overall result being a saving in time and cost, with higher quality. Gaining an adequate *understanding of the system* in the initial stages through "Systems Thinking" will notably reduce the time and cost of the Detailed Design, and the ensuing Production/Integration and Testing, as Figure 1.1 illustrates (Honour, 2004). The striped area affords designers an opportunity to grasp the system's "soul"—what it is all about, what problem(s) it solves, and how it might be implemented.

Honour further contends that "The primary impact of the SE concepts is to reduce risk early" (Honour, 2004). Therefore, rushing off and/or glossing over "System Definition" during the conceptual phase (FEL-1) would increase risks during Detailed Design, Production/Integration and Test. However, an adequate investment in understanding "risks" about a system concept would generally pay off by significantly reducing design risks in ensuing phases.

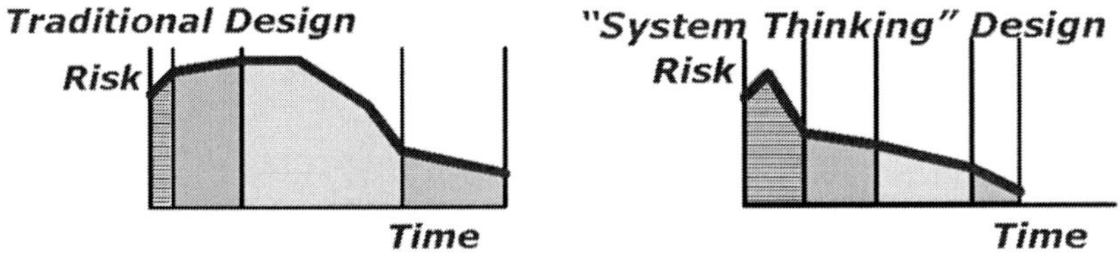

Figure 1.2 – Risk Reduction by "Systems Thinking" Design (Adapted: Honour, 2004)

This approach also makes practical sense, particularly in terms of a system's life cycle costs. Figure 1.3 below shows that up to 70% of the costs of a system are *committed* by activities in the early stage of SE, yet traditional methods seem to be focused on the wrong end of the life cycle. This means the best engineering efforts should be applied in the early stages—as demonstrated, "Systems Thinking helps you think effectively" (Kasser, 2015b). Applying "Systems Thinking" (i.e., gaining adequate insight on the system workings and purposes during the concept phase) not only reduces total delivery costs, but "detecting and rectifying errors" also proves least costly in these early stages.

So Kasser (2007) promotes an "engineering of systems" that begins in the early stages:

> "Back in the 'good old days' of systems engineering, Type III, IV and V [i.e., most senior] systems engineers remedied the problem in the early stage systems engineering activities addressing the conceptual solution. They then produced the matched set of specifications for the implementation or realization of the solution, and moved on to the next contract [i.e., the next system design], leaving the Type II's to continue realizing the solution."

Kasser is not indulging in nostalgia; a flawed concept is likely to breed difficulties later on. This approach requires problem solvers to think in abstract terms in the early stages of identifying a problem and providing a solution (i.e., a concept). Using implementation language (e.g., desired product) in the early stages of problem-solving tends to produce results that may not be the best and/or most innovative solution to the problem, even if it might be a "workable and acceptable" solution at that stage.

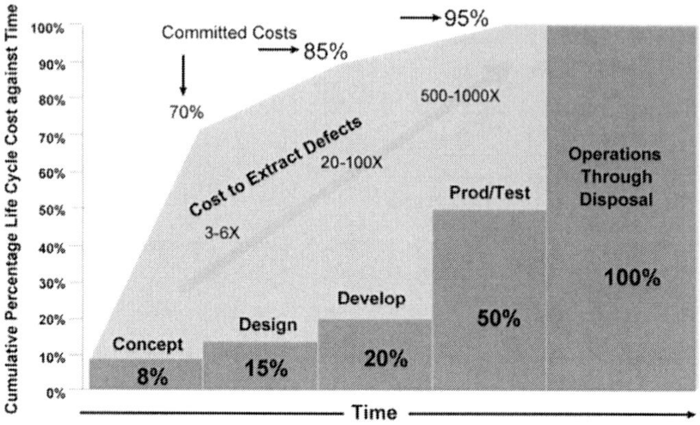

Figure 1.3 – Cost Commitment and System Life Cycle (Fossnes & Forsberg, 2006)

In addition, INCOSE submits that a system is, "A combination of interacting elements organized to achieve one or more stated purposes" (Haskins and Forsberg, 2011). This definition has a useful meaning—the system to be designed ought to fit such a description.

In keeping with this definition, the early focus on exploring the infrastructure as a system (i.e., "*grasping the soul of the new system*") will entail a structured inquiry into what its "components" (i.e., subsystems) might be, their mutual relationships, and how they might interact with one another—i.e., since "*changes in one part may have effects on other parts*" (Bar-Yam, 2014)—and with any adjacent environments (i.e., with pertinent subsystems in such adjacent environments). While it could be premature to expect such an inquiry to afford any detailed information at these early stages, the mere insights that might transpire from this process would definitely bolster the engineering team in their design endeavour.

Furthermore, Systems Engineering (SE) effectively addresses the concern of documenting the Engineering Design and Development (EDD) process and/or "deliverables" as follows:

> "The core element of SE is the explicit documentation of information, something that in the traditional process flow usually only takes place within the heads of those involved. Due to the increased number of internal and external interfaces [i.e., where systems interact] and the fact that information must be transferred between parties and from one phase in the [development and/or] construction process to another, communication of the documented

information is absolutely essential to prevent errors and the cost of failure." (ProRail, 2013)

For instance, the employment of a "Design, Construct and Maintain" approach on the Gateway Upgrade Project in Australia ensured that the project's size, complexity and maintenance requirements were adequately covered. This worked so well not only during project delivery, but throughout the initial ten years of operation of the asset and probably benefited both users and sponsors (Australian Government, 2012). "For all their efforts are to one end: to make something stand that has not stood before, to reassemble Nature into something new, and above all to obviate failure in the effort [...] We must understand the unique engineering problem of designing what has not existed before", Petroski (1992) aptly notes. This *attribute* set humans ahead of other species.

The dictionary definition of engineering is: "The application of scientific principles to practical ends; as the design, construction and operation of efficient and economical structures, equipment, and systems", where the terms "efficient" and "economical" are particular contributions (or features) of good Systems Engineering (Kossiakoff et al, 2011).

There are different degrees of difficulty in engineering design; these levels are categorised as follows:

(i) *Adaptive Design*: In the great majority of instances, the designer's work will be concerned with the adaptation of existing designs. Design activity of this kind demands no special knowledge or skills, and the problems presented are easily solved by a designer with ordinary technical training. For example, the elevator has remained the same technically and conceptually for some time now (Haik and Shahin, 2011).

(ii) *Development Design:* Considerably more scientific training and design ability are needed for development design. The designer starts from an existing design, but the final outcome may differ markedly from the initial product. Examples of this development could be from a car's manual gearbox to an automatic one (Haik and Shahin, 2011), or from oven to microwave—or from elevator to "levitating booth".

(iii) *New Design:* Only a small number of designs are deemed "new" designs when creating a world "as it should be" that is as wonderful as "*The colours of the rainbow so pretty in the sky*" (Mabelo, 2020b) —something that has never existed. This is possibly the most difficult level in that generating a new concept involves mastering

> all the previous skills in addition to creativity and imagination, insight, and foresight. Examples of this are the design of the first airplane or automobile; in fact, even the wheel itself was invented a long time ago (Haik and Shahin, 2011).

When it comes to Large Infrastructure Projects, the type of Engineering Design involved would habitually fall into the first category (i.e., Adaptive Design). Thus, the designer will largely maintain "*the same, technically and conceptually*", merely adjusting the shape, size, and other technical requirements pertaining to the parts, components and assemblies of the envisaged system. As a result, the difference between two random highways could be reduced to a difference in length (i.e., distance from A to B), in width (i.e., number of lanes), and in the makeup and depth of 'sub-structure' materials, as per their load bearing capacity. (A computer will do the trick!)

However, when the complexity of such infrastructure systems increases (e.g., highways with a *draining surface* to avert flooding, coal-fired plants that capture and recycle carbon emissions, railways fitted with monitoring equipment to inform maintenance schedules) there arises a trend to escalate engineering design to higher categories (i.e., Development or New Design).

Accordingly, design engineers may be categorized in four levels as follows: (1) Natural Engineer, who relies on instincts to solve "random problems"; (2) Good Engineer, who indulges in *adaptive designs* and creates "what the customer asks"; (3) Great Engineer, who performs *development designs* and creates "what the customer wants"; and (4) Outstanding Engineer, who is able to produce *new designs* and is able to create "what the customer needs".

One also notes that the complexity of a system is determined by the number of parts or activities, the degree of differentiation between the parts, and the structure of their connections. Complex systems have multiple interacting components whose collective behaviour cannot be simply inferred from the behaviour of the individual components (Gould et al, 1996). Therefore, the complexity of the design would increase with the complexity of the system.

Maqsood similarly argues, "Construction projects are faced with a challenge that must not be underestimated. These projects are increasingly becoming highly competitive, more complex, and difficult to manage. They become problems that are difficult to solve using traditional approaches" (Maqsood et al, 2009). This calls for approaches such as Design Thinking (see Chapter 11) to introduce a new way of "Thinking about Design".

This trend might not necessarily call for a new or more advanced type of design (i.e., convoluted processes), but will possibly require more innovation and, thus, put more emphasis on the need for more frequent, increasingly higher levels of innovation. For example, "Technology continually creates better materials suppression, and detection, which [then] further enhances procedures for the protection of lives and property. Compliance with various codes is not always sufficient, nor are minimum standards (particularly when faced with a fire). Designers should always be aware of innovations to improve mitigation" (Standards and Specifications Specialist, 2018); thus, engineers are enthused to explore innovative methods (NJDOT, 2016)—and many SE experts concur:

> Kasser — "Systems thinking alone cannot help you provide innovative solutions to complex problems. This is because understanding situations [through systems thinking] is only the first step on the journey that provides those innovative solutions." (Kasser, 2015)

> Kossiakoff — "Just as important, however, is a side of systems engineering that is often overlooked, namely, innovation. For a new complex system to compete successfully in a climate of rapid technological change and to retain its edge for many years of useful life, its key components must use some of the latest technological advances. These will inevitably introduce risks, some known and others as yet unknown, which in turn will entail a significant development effort to bring each new design approach to maturity and later to validate the use of these designs in system components." (Kossiakoff et al, 2011)

> Forsberg — "The conventional mode of focusing the team just on risks tends to foster negativism. An alternative is to have the [project] team seeking and seizing opportunities to excel and then to examine and manage the risks of those opportunities. This approach encourages innovation and fosters positive teamwork [for EDD]." (Forsberg et al, 2005)

> Eduardo — "Thus, the project is the primary unit for making business and promotes development and innovation. And it has multidisciplinary teams [including engineering experts] to deliver complete solutions to customers, cutting across units' boundaries." (Eduardo et al, 2000)

> Jamshidi — "Here, innovation includes both technological innovation and organizational and human conceptual innovation [...] Human ingenuity and technological innovations in recent years have led to rapid advancements in the emergent [and futuristic] fields of robotics, mechatronics, information, and computerized systems. This has created a new type of system of systems that may be grouped under the category of intelligent systems [such as *integrated* transportation systems]." (Jamshidi, 2009)

CHAPTER 2

THE ENGINEERING SPECIFICATION PROCESS

Traditionally, whenever a need arises to devise a new system, component or process to meet the desired needs of a customer, "a formal design team may be assembled to plan the physical proceedings, and to integrate those proceedings with the other parts. The design usually consists of drawings and specifications" (Hussin and Omran, 2009). However, good practice in Large Infrastructure Projects will generally entail many other artefacts, deliverables and such items. "To complicate matters further", Haik and Shahin (2011) complain, "everyone has a different opinion/desire on how a product [or process] should be designed". Thus, Engineering Design and Development (EDD) needs therapy.

Two maladies do plague engineering endeavours in projects, namely (1) the propensity of design engineers, being humans, to still make mistakes, and (2) the implicit and discretionary approach to EDD in traditional methodologies, which makes the whole exercise *inconsistent* and, therefore, difficult to manage. The author takes the view that a set of structured, robust, and above all *systemic* processes (i.e., based on Systems Thinking and Systems Engineering) would sufficiently address/remedy both concerns.

Haik and Shahin (2011) also maintain that there are various forms of the systematic design process, and different sources list as few as four to as many as nine design processes. Essentially though, they all revolve around the same "generic" design stages as follows:

(i) Requirements—i.e., identifying customer needs, carrying out a market analysis to locate what is already available in the market, and defining design goals. "Many of the difficulties encountered in design may be traced to poorly stated goals [i.e., poor requirements] or goals that were hastily written and resulted in confusion [i.e., wrong requirements] or too much flexibility" (Haik and Shahin, 2011).

(ii) Product Concept—i.e., for establishing functions and tasks specifications. "It is also useful to identify the functions instead of the potential solutions. This is sometimes referred to as remaining 'solution neutral' (i.e., no solution is offered at this stage) [but only the required functions]" (Haik and Shahin, 2011). It starts with a Concept of Operations (ConOps).

(iii) Solution Concept—i.e., the conceptualisation and evaluation of alternatives. "This activity usually requires *free-hand sketches* for producing a series of alternative solutions. The alternatives do not need to be worked out in detail but are recorded as possibilities to be tested" (Haik and Shahin, 2011). Those (functions-) alternatives shall transpire from confirmed ConOps and should be listed in an organised fashion.

(iv) Embodiment Design—i.e., this stage will not include any (engineering) details yet, but will begin to illustrate a clear definition of a part, how it will work, and how it interfaces with the rest of the parts in the product assembly. This stage will therefore include considerations as to "How to build" the whole product (i.e., constructability).

(v) Detailed Design—i.e., this final/ultimate stage includes analysis and optimisation of dimensions and tolerances of parts, and their testing. The dimensions of shape, size and technical characters are derived from specified loading (Haik and Shahin, 2011).

The majority of design engineers are mostly conversant with Detailed Design. This fact alone explains to a large extent the concerns around traditional approaches to engineering design, i.e., the propensity for errors and inconsistent processes. Haik and Shahin (2011) suggest that such an infirmity arises from training (i.e., revolving around detailed design):

> "Detailed Design [stage] […] is what most of the engineering courses in an undergraduate degree program cover. In essence, the solution must be tested against the physical laws. [But] [t]he manufacturability of the chosen product also must be checked to ensure its usefulness. A product may satisfy the physical laws, but if it cannot be manufactured, it is a useless product."

However, Detailed Design is in fact increasingly being executed by computer-aided tools. The Systems Thinking approach, in contrast, will recommend an approach that "*starts with the end [i.e., operations] in mind*", which is an SE principle (Scott, 2012). Therefore, the author has made this important point in many of his previous publications as follows:

> "Moreover, one important implication of applying Systems Engineering concepts and principles to enhance project life cycle methodologies entails a life cycle that exhibits the Operational Environment [i.e., beyond Project

> Closeout], with the [new] understanding that projects are primarily about improving (or establishing) that particular environment. The whole idea is that the entire project team starts the project with the end in mind." (Mabelo, 2016; 2020; 2020b)

Thus, Engineering Design should seek to devise a system and/or process that improves (or establishes) the operational environment—to improve the business (Mabelo, 2020b).

The author upholds that, "The effective identification of stakeholders and, among others, their requirements (e.g., Stakeholder Requirements) shall flow from the 'Concept of Operations' or System ConOps, which is viewed as part (i.e., the first step) of the SE life cycle process. The ConOps [i.e., "*A day in the life of the system*" (Mabelo, 2020b)] also defines the system's goal (i.e., the *Why*) towards the achievement of which stakeholders shall be managed and, as such, it also informs who the stakeholders are" (Mabelo, 2020).

Moreover, customers (and users) are among the most important stakeholders in Large Infrastructure Projects. Thus, Koskela and Howell (2002) submit, "In the value generation view, the basic thrust is to reach the best possible value from the point of the customer". They further maintain that, "The value generation view provides for an explanation on the third goal of project management, delivering the business purpose [or goals]. Principles related to rigorous requirement analysis and systematized flowdown of requirements, for example, are forwarded" (Koskela and Howell, 2002).

It is quite unfortunate that customer requirements are not always taken into account in infrastructure projects. This arrogant stance on the part of engineers creates problems. Typically, when customer requirements are poorly investigated or ignored at the onset, the process of requirements clarification and change leads to disruption in the progress of the project—and will cause overruns. The actual progress starts to drift from the plan, the updating of which is too cumbersome to be done regularly (Koskela and Howell, 2002).

One such arrogant attitude might stem from "*design by force of habit*"; engineers prefer designs similar to what they have enjoyed before. During a mega project the author was involved in, an intense debate once arose between the civil works project manager (from South Africa) and the engineering manager (from Canada) concerning the type of sewage treatment plant to be designed and installed. The latter had recommended a *world-class* design, claiming he would not specify "*anything lower than what he has in his backyard*", whereas the former was insisting on a design that could be procured from, installed and repaired by South African companies, "*just next door*". The host country (Mozambique), their preferences, and their technological circumstances hardly featured in those disputes.

Some 12 years later, in the same country, a major new bridge was found to be *over-designed* and, as a result, the provincial infrastructure agency lacked the ability to maintain the intricate system; its high-tech monitoring equipment was also stolen. "Thus, customer requirements cannot be taken as given and unproblematic" (Koskela and Howell, 2002) or else the project would not have begun with the end in mind, resulting in "reduced efficiency and lessened value to the customer" (Koskela and Howell, 2002). Surely, operations were not improved in this instance—a lesson we ought to learn.

Therefore, "The most important step of the design process is identifying the needs of the customer at the 'requirements' stage. However, before this is done, it is important to establish who the customers [i.e., stakeholders] are" (Haik and Shahin, 2011). Further, the starting point for a Specification and Design Process is the problem definition. Quite sadly, however, "Too many times, individuals and teams jump into problem-solving activities without fully or properly defining what it is they really need to solve, or what factors or interactions [with any 'adjacent' systems] within the problem area will create complications" (Borza, 2011).

Has not Ackoff warned us all that, "The *righter* we do the wrong thing, the *wronger* we become"? (Ackoff, 2004). Surely, "It doesn't help to solve the *wrong* problem?"—Law number #2 of Effective Systems Engineering, as articulated by Zane Scott (Scott, 2012).

Organisations would become *wronger* whenever their large infrastructure projects fail, or whenever they deliver a somewhat "*square peg in a round hole*" kind of infrastructure system. The fact is, such fatal predicaments start from a failure to manage requirements. Therefore, the inadequate and/or ineffective management of project requirements more often than not result in difficulties or failure.

> "A factor [that is] present in every successful project and absent in every unsuccessful project is sufficient attention to [the "holistic" management of] requirements." (Robertson and Robertson, 2005)

Requirements engineering and management greatly enhance the success of projects. Hence, it is vital to understand the importance of successfully identifying and managing product/process requirements. SE concepts are used to improve project success through better handling of requirements, thus, reducing "scope creep" and unwanted surprises, e.g., delivering the wrong solution (thus, making project outcomes and/or operations *wronger*).

In the unfortunate event (but highly likely, in large and complex infrastructure projects) that the requirements delivered during development are not of a high quality and/or relevancy, the solution derived from such requirements runs the risk of not achieving what it is intended for (i.e., the flawed, wrong, or unintended solution). Sometimes, faulty assumptions or errors taint requirements, which then cascade through the basis of design (Merrow, 2011). It is *cheaper* to fix system errors at requirements level —by fixing their quality and/or relevancy.

The author previously discussed a Project Failure Grid with two axes (Mabelo, 2016). The *traditional* Delivery Performance axis (i.e., on time, on budget, to specification) and the Business Improvement axis (i.e., improved operations, strategic goals) both range from "low" to "high". This grid indicates four quadrants pertaining to project outcomes:

1. Total Success—whereby both delivery and business targets are satisfactorily met
2. Delivery Failure—whereby delivery targets are missed, but business targets met
3. Business Failure—whereby delivery targets are met, but business targets missed
4. Catastrophe—whereby the project has failed on both delivery and business axes, having been aborted/canned with nothing achieved to show for the resources spent

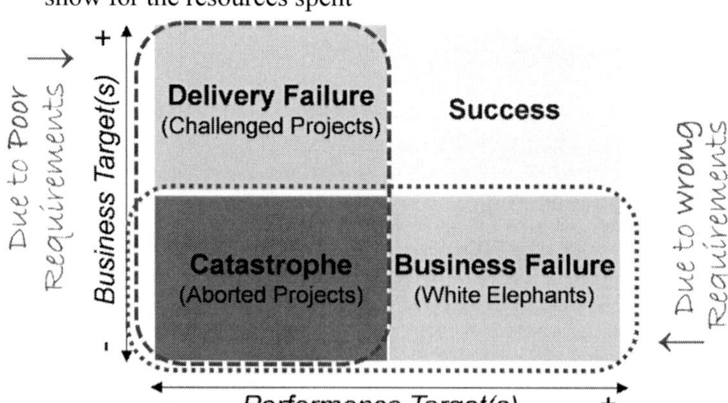

Figure 2.1 – Requirements and Project Failure Grid (Mabelo, 2016)

Figure 2.1 juxtaposes requirements against the Project Failure Grid to show that "poor" requirements (i.e., characterised by low quality of statements) are likely to cause delivery failure. In the same way, "wrong" requirements (i.e., characterised by a lack of relevancy in statements) are likely to result in business failure. Worse still, should requirements prove both "poor" and "wrong", they are highly likely to produce (a string of) aborted projects.

To understand what "poor" (i.e., condition or capability expressed in ambiguous or unclear manner) and "wrong" (i.e., expressing irrelevant condition or capability) actually mean, one shall recall the IEEE 610 (1990) definition of a "requirement" (Kasser, 2015):

> *"(1) A condition or capability needed by a user to solve a problem or achieve an objective.*
> *(2) A condition or capability that must be met or possessed by a system or system component to satisfy a contract, standard, specification, or other formally imposed documents.*
> *(3) A documented representation of a condition or capability as in (1) or (2)".*

However, many project-entities still fail to produce satisfactory requirement statements. "The failure to complete FEL [i.e., develop business case, as well as *translate* customers' needs into requirements, then into design solutions] on megaprojects is costing industrial firms and their shareholders hundreds of billions of dollars of value" (Merrow, 2011). Thus, it is advised that, "After determining what items are most important to the customer [i.e., the voice of the customer], organizations must translate them into particular specifications [i.e., sets of requirements]. Nothing [new] can be produced, serviced or maintained without detailed specifications or some set of given standards. Each aspect of the desired item must be clearly defined" (Tapke et al, 2001). The reality, particularly in LIPs, is still far from here:

> "Most of the time the customer provides a generic statement of need, and it is up to the engineer to identify the specific needs of the customer. For example, we are required to design a chair that can be used by a child. Clearly, all of us know how to sit on a chair, so in that perspective, we know how a chair works. A chair is used for sitting. <u>Unfortunately, this description does not say how the chair is made. What material is used? Is the chair flexible or rigid? Does the chair rotate or is it fixed? [Is it collapsible?] What does it mean that the chair is to be used by a child? Is safety the biggest concern? How much will the chair cost? How old is the child? And so on</u>." (Haik and Shahin, 2011). [emphasis added by the author]

Until such details are adequately elicited, there are no requirements—very little chance of success, no matter are *committed* the project team might be!

On a project involving a new library, the author subjected the "Bathroom Requirements" to a quick review process using a patented software to ascertain their quality status. The outcome confirmed his concerns that the "nebulous" requirements put out on tender will invite flawed/misguided responses from most bidders, and confusion and disputes later on. The requirements were so *poor* (i.e., in terms of quality) the project was heading straight to overruns, if not failure. The requirements review report indicated a number of "flaws in requirements" as follows:

- 95% of requirements entries were categorised "Very high risk" (i.e., requirement lacks an imperative, or contains multiple imperatives, and/or contains excessive vague terms – Consider rewriting the requirement or separating into multiple concise requirements)
- 36% of items included "vague words" (e.g., *Ensure that correct water level has been set*)
- 27% included "excessive continuances" (e.g., *Operating overflow tube height conforms to and operates to local bylaws and SABS specification*) – How do you even verify this?
- 9% involved "universal quantifiers" (e.g., *Complete all plumbing*; *Always use silicone*)

"The purpose of Requirements Engineering, a discipline at the heart of SE, is to maximise the likelihood that a developed or bought solution or maintenance initiative will deliver solutions that function as desired" (Marnewick et al, 2014). Thus, the customer ought to allocate a budget of money and time (e.g., up to 40% of the development time in ICT, but probably far less in infrastructure) in order to produce a good requirements specification.

The highest level of requirements is the "Undesirable Situation", the primary problem (Kasser, 2015). The "solution to the primary problem" constitutes the objective for the particular project; therefore, this is the primary objective of the EDD endeavour. For instance, since the morning traffic congestion from Pretoria to Johannesburg was costly to the South African economy, a system and process (i.e., "Gautrain" rapid train) has been devised to address that predicament. This primary problem statement is, thus, referred to as the project's primary requirement and is usually translated into a number of concepts of operations (ConOps) to determine the stakeholders and their requirements. This provides project teams with an "end-image", a target to plan and design towards (i.e., "Design to").

SE subsequently proceeds on the basis of two processes running in parallel to achieve that very objective, namely (1) the Specification Process and (2) the Design Process (ProRail, 2013). Furthermore, there is a growing trend to further formalise the design process, leaning more towards addressing the problem, and postponing the solution to the later stages, rather than finding a solution early on and then trying to improve it (Haik and Shahin, 2011).

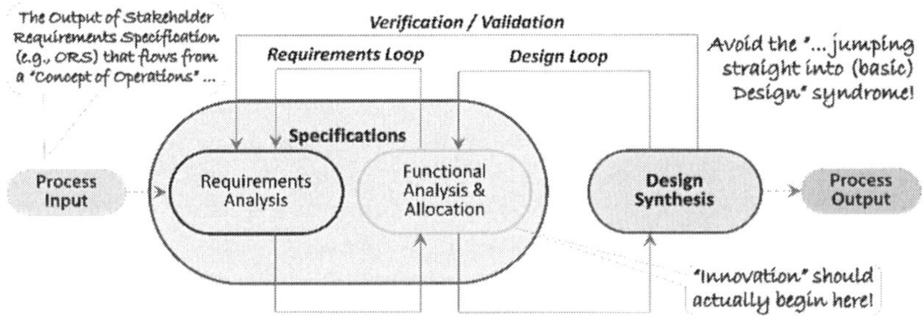

Figure 2.2 – Engineering Specification Process (Adapted: ProRail, 2013)

This is a welcome departure from the traditional (yet still prevalent) urge to jump into Detailed Design. Unless a "process" is in place, designers will tend to jump to Detailed Design and deprive themselves of opportunities (i.e., reviews) to innovate and detect/rectify their errors. Further, they may not be able to establish what "Tasks, resources, and risks" are involved.

Sullivan (an architect by profession) made an insightful statement. Although he was talking about "buildings", his dictum would apply to any complex system. In designing a system (any new system) the designer should *first* consider the functions that the system or component shall fulfil before exploring its form (i.e., shape, size) and performance — "*Form ever follows function*", Sullivan (1896) aptly noted. Besides, in exploring a number of "combinations of functions" designers would more often than not attain innovation in the design endeavour.

A similar, iterative design process (see Figure 2.3 below) is promoted by Kossiakoff who describes a structured and "systemic pathway" from Needs Analysis (either Operational Deficiencies or Technological Opportunities) to Concept Exploration (i.e., System Performance Requirements, Candidate System Concepts) and finally to Concept Definition (i.e., System Functional Specifications, Defined System Concepts) (Kossiakoff et al, 2011).

No matter which of these two processes design teams might elect to apply, the wise words of Plato still ring true as a stern warning to designers: "*The beginning [i.e., primary problem] is the most important part of the work*".

In the unfortunate, but likely event of the system designer erring in the problem definition (e.g., misdiagnosis of the primary problem, solving the wrong problem), the engineering team would find it *hard* translating requirements into functions and allocating subsystems (function enablers) in ways that produce the right improvements in operations. The solution will not fit—Fixes that fail!

In line with the very essence of Systems Thinking, there are feedback mechanisms (i.e., loops, to be discussed hereafter) to allow for adaptation. Jamshidi (2009) also notes, "Similar to living organisms, an engineered SoS [System of Systems] can be built to have conformer, regulator, and negative as well as positive feedback control mechanisms […] The feedback loops […] are within the context of systems engineering; they serve to refine the analysis and modeling steps […] This repetition provides a feedback mechanism [i.e. loop] that compensates for model error [e.g., the Specification Process resulting in wrong functions and then subsystems being chosen]".

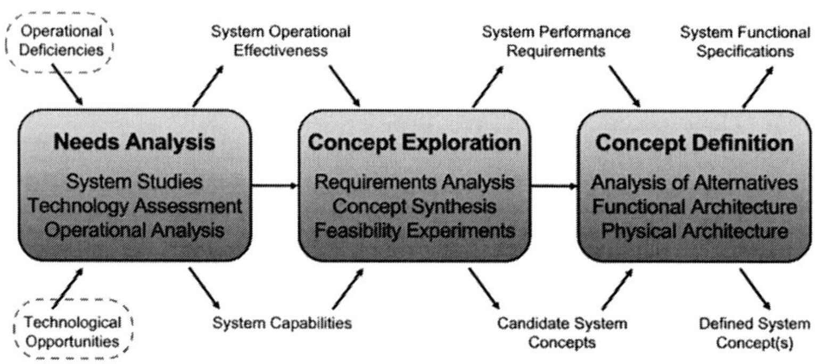

Figure 2.3 – Engineering Concept and Specifications (Source: Kossiakoff, 2011)

Whether it applies to simple buildings or to complex systems such as bridges or power plants, "function" (i.e., Functional Behaviours) ought to come *first* and inform/determine the physical architecture (which is part of the "form" realm). The practical implication of this principle is that in designing systems (e.g., LIPs), Functional Analysis and Allocation should come before and inform Detailed Design. The design process (Figure 2.2 above) as recommended by the Dutch Ministry of Infrastructure and the Environment (MIE) involves two separate but crucial processes (ProRail, 2013); jointly, they constitute the Specifications Process as follows:

(1) The Requirements Analysis Process:

Its objective is to translate stakeholders' requirements (or needs) into measurable system requirements and required functions. This process generally includes Systems Requirements, Functional Requirements, Functional Limitations, Interface Requirements, and Non-Functional Requirements (Table 2.1 seeks to define some of these SE concepts). During this Requirements Analysis stage, logical models are used that will describe "What the system shall do" in order to meet its purposes.

These wishes and needs of relevant stakeholders are often expressed as required "functions" that the system will have to perform during its life cycle—functions being "actions, tasks or activities" that the system is to perform. During the Requirements Analysis process, these wishes and needs are analysed and then weighted in order to arrive at a set of requirements. Hence, the typical activities to carry out during this stage would generally include the following:

(i) Analysing the "desired/projected" needs for the new system, e.g., superior performance, enhanced interoperability, safer usage, or ease of maintenance
(ii) Understanding the "value" of fulfilling the wishes/needs of key stakeholders throughout the system life cycle. (Here is a good place to commence Value Engineering/Management, not after the detailed design is complete!)
(iii) Defining the applicable quantitative operational/performance objectives, as well as exploring plausible and pertinent Concept of Operations (ConOps).

These (and other similar) technical processes are required to create and use a system, and they apply at any level in a "Hierarchy of Systems" structure (for one systems engineer's subsystem is another systems engineer's system). Jamshidi (2009) notes, "The technical processes [of SE] are concerned with technical actions throughout the life cycle. They transform the needs of stakeholders first into a product and then, by applying that product [i.e., putting the new infrastructure into operation], provide a sustainable service when and where needed to achieve customer satisfaction".

It evidently follows that should the needs of the stakeholders not be taken into account *first*, the resultant product/system would fail to satisfy the customers (Mabelo, 2020). Engineers shall therefore depart from a traditional product-driven approach to LIPs design to embrace a customer-driven approach. For "The most important issue in project management is for the project manager to get project [and engineering] staff, beneficiaries [i.e., customers], and other stakeholders to develop a common understanding, agreement, and commitment to a project's objectives" (Youker, 1998).

SE Concept	General Description/Definition
System	It is a combination of "interacting" elements (not a mere group) organised to achieve one or more stated purposes.
Subsystem	▪ By their nature, complex systems have a hierarchical structure in that they consist of a number of major interacting elements, generally called subsystems. ▪ It is a set of elements that constitutes a system and performs desired operational functions (or *behaviours*).
Concept of Operations (or ConOps)	The Concept of Operations describes how the system will be used/operated during the life cycle phases to meet the expectations of relevant stakeholders and thus helps facilitate a "holistic" understanding of the system goals.
Systems Requirements	As the *totality* of requirements, they result from analysis of the input Stakeholder Requirements; they are required to meet project and design constraints such as Functional Requirements, Non-Functional or Quality Requirements, Performance Requirements or Architectural Constraints.
Functional Requirements	They largely refer to what the system should do, and they should be "action" oriented and describe the tasks or activities that the system performs during its operation. E.g., read fingerprints; emit "grey" light; stop water flow; or produce 250 litres of drinkable water per person/day.
Interface Requirements	The functional, performance, electrical, environmental, human, and physical requirements and constraints that exist at a common "boundary" between two or more functions, subsystem elements, systems or configuration items the system shall be devised to accommodate. E.g., transfer data to YZ mobile device; discharge no/little CO_2 in atmosphere.
Non-Functional (or Quality) Requirements	These are used to evaluate the system performance, i.e., performances, operational conditions and constraints. E.g., operate under sea; be available 24-7; have cleanable casing.
Functional Architecture	An architectural model that identifies system functions and their interactions. It defines how the functions will operate

	together to perform the system mission(s). In general, more than one architecture can be found that fulfils requirements.
Functional Limitations	These constraints recognise inherent limitations on the shape, sizing or capabilities of the system, its interfacing systems, and its operational and physical environment.

Table 2.1 – Systems Engineering Concepts and General Definitions

Regrettably, many believe customer requirements are not really relevant to LIPs. Those engineers still believe, "*The market [i.e., group of potential customers] does not know what they want; it is the engineer's job to tell them*"—tell them "*What it is*" that they want; "*You want a railway, we will give you one; a reservoir, you shall get one*". When was an architect or town planner last involved in designing a water reservoir? Engineers give little thought as to how customers may feel about its lack of visual appeal, or how it might even blemish the city's skyline. Infrastructure requirements should include aesthetic functions.

Moreover, customers and other stakeholders will usually express their needs in terms of their daily experience, not necessarily what their *real* issues are. Thus, there still remains the necessity of *figuring out* what the real needs of customers and stakeholders might be.

Nevertheless, Systems Thinking would maintain that, "Because 'systemic structures' generate 'patterns' and 'events'—but are very difficult to see—we can imagine these three levels as a kind of iceberg, of which 'events' are only the tip. Because we only see the tip of the iceberg, the 'events', we often let those drive our decision-making. In reality, however, the 'events' [i.e., occurrences we usually encounter in operations] are the results of deeper 'patterns' [i.e., accumulated memories of events and trends] and 'systemic structures' [i.e., ways in which a system's components are arranged]" (Kim, 1999).

Most customers will only perceive the "events"—as the "late bus" scenario involving the damsel in Figure 2.4 may suggest. The father, probably a trained systems engineer, is more concerned about systemic issues; for instance, buying her a car might not help as long as traffic lights on the way to and from the mall are faulty—and had the mall been situated at some walking distance from their home, no car would be needed by the beloved daughter at all!

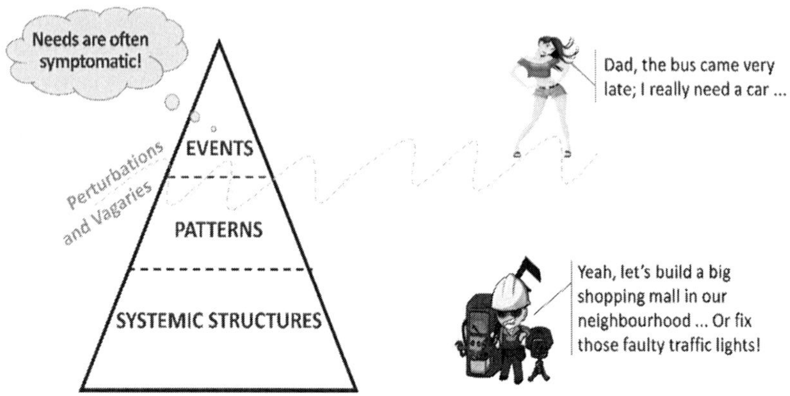

Figure 2.4 – Three Levels of the Systems Thinking Iceberg

Accordingly, devising infrastructure systems based on "events" will more often than not result in interventions that only address symptoms (i.e., fixes that fail). Engineering teams ought to dive beneath the "perturbations and vagaries" of daily operations, seeking to not only perceive the trends and patterns of undesirable situations, but to further comprehend their "systemic structures", or the way in which the system components (e.g., subsystems) are configured. In so doing, engineers will create solutions "*for both now and the future*". For it is the systemic "architecture" that governs the overall system's "behaviours", as well as "evolution" (due to external interactions).

Hence, the engineering intervention should seek to "*identify the design components or modules (e.g., subsystems) that must be added, changed or modified for each identified need*" (see Chapter 11). When this goes amiss, it can drag an organisation or a project down even more quickly than the initial problem. Indeed, as Crouhy argues, "A risk management strategy [in terms of engineering design errors and implementation mistakes] that is not carefully structured and monitored is a double-edged sword: if it goes wrong, it can drag a firm [or any large project] down even more quickly than the underlying risk" (Crouhy et al, 2009).

(2) The Functional Analysis and Allocation Process:

Its objective is to transform the required operational functions of a system into "subsystems" (often referred to as "Function Enablers" or "Behaviour Enablers") and to prepare a specification that documents the requirements that these selected subsystems are expected to meet. Subsystems are at the heart of the design; "Axiomatic design [...] advances further the principles along which [customer] requirements should be assigned to product subsystems [thus, its architecture], a significant issue of value generation [...] [It] admits that at the outset, customer requirements are not necessarily available or well understood, and that the allocation of requirements to different parts of the (project) product is a difficult problem" (Koskela and Howell, 2002).

Henri Poincaré (1854–1912), the famous French inventor, certainly encourages such a practical (yet insightful) viewpoint regarding the "Functional Definition" undertaking as follows:

> "To create [a system] consists of making new combinations […] The most fertile will often be those formed of elements [i.e., subsystems] drawn from [engineering] domains which are far apart [...] To create consists precisely in not making useless combinations and in making those which are useful and which are only a small minority. Invention is discernment, choice." (Henri Poincaré, cited by Popova, 2020)

This applies to infrastructure projects too, particularly owing to their ever-increasing complexity and importance in our life. Our modern society relies on infrastructure to maintain a decent standard of living.

Choosing the most fertile "*combination of subsystems*" is crucial to system design. The last thing one expects is a new system that cannot perform the desired functions. Haik and Shahin (2011) would agree with this stance: "In reality, the designer is trying to assess what actions [or functional behaviours] the product [as a system] should perform during its lifetime and operation. This technique allows for alternatives to be explored [i.e., *optioneering*] that can address the needs and goals rather than fixating on a solution that the customer provides unintentionally [...] This stage of the design [...] guides the designer to a problem-focused design rather than a solution-focused one".

It is no wonder most of the innovation would take place during this *vital* process of Functional Analysis and Allocation. Jamshidi (2009) suggests the envisioned innovation could (or rather should) be "motivated by four innovation drivers (i.e., collaboration, customization, integration, and adaptation). Not surprisingly, all four drivers are directed at empowering the individual, that is, at recognizing that the individual can, respectively, contribute [positively] in a collaborative situation, receive customized or personalized attention, access an integrated SoS, and obtain adaptive real-time or just-in-time input".

Jamshidi goes on to argue that, "*Innovation includes both technological innovation and organizational and human conceptual innovation*". Thus, tools and approaches such as Design Thinking would assist in fostering collaboration, customisation, integration, and adaptation of new ideas. Haik and Shahin then provide a pertinent example of Functional Analysis that highlights the significance thereof:

> "A client may ask for a traffic light system to be placed on a particular junction, where in fact, an underpass may be a more viable solution to achieve the real goals of the task [or the project], which is to alleviate traffic congestion." (Haik and Shahin, 2011)

Functional Analysis is too crucial to the EDD process to be omitted or compromised.

Haik and Shahin (2011) also encourage accommodating evolution in innovation. "Proficient [system] designers control evolution and innovation so they occur simultaneously. Although the emphasis is on innovation, designers must test their ideas against prior design. Engineers can design [systems] for the future but must base results on the past". The newer shall supersede the older, just as it was said: "*The elder shall serve [i.e., clear the way for] the younger*" (Bible: Genesis 25:23).

Maier and Rechtin (2009) summarise the essence of the innovative and credible design of SE (and source of risks) with regards to "system interfaces" as follows:

(i) "Relationships [i.e., not merely the individual strengths] among the elements [of the system under design] are what give systems their added value."
(ii) "Choosing the appropriate aggregation of functions [as part of the Functional Allocation] is critical in the design of systems […] in partitioning, choose the so that they are as independent as possible, that is, elements with elements low external complexity

and high internal complexity [for example, far higher complexity in the engine than in its associated drive shaft] […] choose a configuration in which local activity is 'high speed' and global activity is 'slow change' […] [and which entails] minimal communications between subsystems." In this principle lies the *secret* of a smart system architecture.

(iii) "The greatest leverage [i.e., value creation] in system architecture and design is at the interfaces [i.e., where two or more systems and/or subsystems interact]."

The Functional Analysis and Allocation Process, which should precede and, more importantly, inform the design should involve the following steps (ProRail, 2013):

(1) Provide detailed specification of all of the system's functions (i.e., behaviours)
(2) Derive the subsystems (i.e., Function Enablers) from these functions
(3) Create innovative structure and coherence among these subsystems
(4) Link the requirements from Requirements Analysis phase to the subsystems

These functions are *iteratively* transformed into "solution-independent" subsystems, considering that a particular subsystem may perform multiple functions (e.g., a jet-engine that can also brake, an on-off switch that can also manage sound volume). The result is an overview of all requirements, as well as the interfaces known at that stage, and a specification of each subsystem as a basis for ensuing detailed designs.

The Dutch MIE has published an example of Functional Analysis and Allocation for a "Water Weir and Lock Complex" (see Figure 2.5 or the author's example in Figure 9.2). The owner's needs/policies are translated into Functional Requirements (and Derived Functions) which are suitably allocated to relevant subsystems, including the 'Weir', 'Lock', 'Information and Energy', 'Building', as well as 'Ground' subsystems that are to perform such functions as allocated—they constitute the "System". Kossiakoff notes that, "By their nature, complex systems have a hierarchical structure in that they consist of a number of major interacting elements, generally called *subsystems*, which themselves are composed of more simple functional entities, and so on down to primitive elements such as gears, transformers or light bulbs, usually referred to as *parts*" (Kossiakoff et al, 2011).

In accordance with such a *systemic* perspective, Kossiakoff will further maintain that:

> "The [SE] approach tends to be more technical, seeking from potential future users and developers of the solution system, what are the top level needs, requirements, and concepts of operations [i.e., ConOps], before conducting a functional and physical [or detailed] design, development of design specifications, production, and testing of a system solution for the problem. Attention is given to the subsystem interfaces and the need for viable and tangible results. The approach and practical end could be applied to many degrees of complexity, but there is an expectation of a successful field operation of a product [or system and/or process]. The proven reliability of the systems engineering approach for product development is evident." (Kossiakoff et al, 2011)

He also argues that, "Another interpretation of systems engineering comes from engineers who develop [...] who lean heavily on the systems engineering principles that focus on management of interfaces and feedback systems [...] a type of feedback that is characteristic of the system development process" (Kossiakoff et al, 2011).

Thus, the Specifications Process (see Figure 2.2) is also typified by a feedback loop linking its two main elements: Requirements Analysis and Functional Allocation. This Requirements Loop plays a systemic role as the Dutch Ministry of Infrastructure and environment (MIE) stipulates:

> "The Functional Analysis may/should provide insight into points where the Requirements Analysis process was lacking. Thus, the Requirements Loop then returns to the [prior] Requirements Analysis and the analysis is repeated, transforming Requirements Analysis into an *iterative* [but still structured] process that should be repeated several times" (ProRail, 2013).

This is in accordance with the cybernetic perspective of SE. Cybernetic control is the ability to predict and control a system's behaviour through proactive (i.e., feedforward loop), as well as reactive (i.e., feedback loop) mechanisms to process information, react to information, and evolve to adapt. And Jamshidi (2009) provides a practical example:

> "Consider the application of monitoring a large structure such as a bridge. Using sensor networks [...] requires sustaining the lifetime of the deployed sensors for a long time, since the redeployment generally can be difficult, in terms of both ease and cost of deployment."

This feedback loop (i.e., Requirements Loop) provides a potent enhancement to design. For example, "A human can tentatively imagine the main features [i.e., requirements] of an advanced transportation vehicle having certain capabilities for satisfying transportation objectives, which can then be rigorously analysed [i.e., Functional Allocation] and viewed graphically using a Computer Aided Design (CAD) [i.e., for the "What to build"] and/or Manufacturing (CAM) program [i.e., for the "How to build"]" (Jamshidi, 2009). Jamshidi further argues, "Based on this analytical and visual feedback, the vehicle can be redesigned and analysed again in an iterative manner until a satisfactory design is achieved that meets specified performance (e.g., human interface, environmental, fuel efficiency) criteria" (Jamshidi, 2009)—feedback is appreciated in Systems-of-Systems (SoS):

> "Since SoS are often fielded suites of systems, feedback on SoS performance may be based on operational experience and issues arising from operational settings. By monitoring performance in the field or in exercise settings, areas for attention [e.g., to improve the design] can be identified and impacts of unplanned change in constituent systems can be assessed." (Jamshidi, 2009)

Further, Testing and Evaluation (T&E) are inherent (and critical) functions of engineering design. "The function of T&E is performed by the artist as part of the process of transferring a design concept to canvas. To the extent that the painting does not conform to the artist's intent, he or she alters the picture by adding a few brushstrokes" (Kossiakoff et al, 2011). The design is a "closed-loop" process in which T&E constitute the feedback which adjusts the result to the requirements (at relevant system levels) that it is mostly intended to meet. Kossiakoff provides further guidance as follows:

> "Systems engineering is primarily concerned with guiding the engineering development and design, defining and managing interfaces, developing test plans, and determining how discrepancies in system performance uncovered during test and evaluation (T&E) should best be rectified." (Kossiakoff et al, 2011)

> "The simulation of the system operational environment for purposes of system test and evaluation can be a task of major proportions, sometimes approaching the magnitude of the system design and engineering effort itself." (Kossiakoff et al, 2011)

This issue is critical in Large Infrastructure Projects when construed as SoS:

> "For an SoS, double-loop learning challenges error in ways that call into question the ways an SoS has been [initially] designed and is operating. This is especially important where there are high levels of emergence and rapidly shifting contexts. What was once essential [e.g., key system requirements] may no longer be appropriate for an SoS." (Jamshidi, 2009)

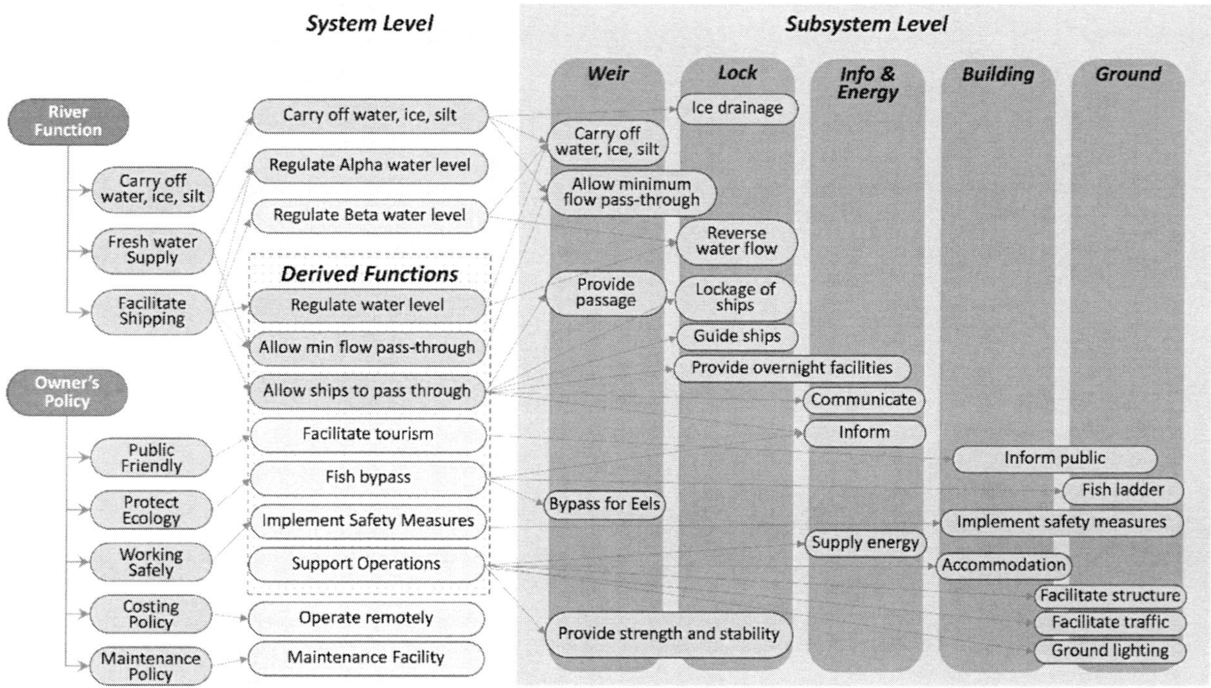

Figure 2.5 – Example of Functional Analysis and Allocation (as per ProRail, 2013)

CHAPTER 3

DESIGN AND DEVELOPMENT: THE PROCESS

During the Design Process, the selected subsystems are developed in accordance with the previous Functional Analysis. At this phase, a "solution-independent" subsystem (i.e., based on Functional Behaviours) is transformed into a physical solution-based subsystem, which then informs the System Architecture—thus, the overall behaviour of the system.

Following an exploration of the "combinations of functions" and the ensuing selection of the *pertinent* subsystems (those that enable the system to attain its intended performance), the Design Process produces the "design choices" that best meet the client's requests and objectives, and therefore provides the transition from problem to solution. Such a design (rather the resultant system and/or process) must not only meet the functional requirements, but also the various aspects-related requirements throughout the life cycle — "*from womb to tomb*" as the author puts it. Requirements should, thus, involve maintainability, ease of inspection, and disposability.

Regarding the transition from "function selection" stage to the subsequent "detailed design", Ryschkewitsch et al (2009) consider the Functional Allocation and Architecture as follows:

(I) Architecture encompasses the fundamental organisation of a system, embodied in its components, their relationships to each other and the environment, and the principles governing its design and evolution. This Ryschkewitsch definition of "Architecture" seems to conflate Architecture and Behaviours, a stance Scott (2014) frowns upon.

(II) Design is about creating a product or system, as well as a plan to develop and use (or operate) it. For our purposes, (system) architects provide the rules; (system) designers create implementations using those rules (Ryschkewitsch et al, 2009).

(III) Systems engineers (e.g., in large infrastructure) do both; they help create the design and maintain its integrity throughout the life cycle (Ryschkewitsch et al, 2009).

In addition, it should be noted that implicit in the "physical definition" step is the definition and control of interfaces, both internal and external. Each element (i.e., subsystem, assembly, part) added or elaborated in the Design Process should be properly connected (i.e., effectively interfaced) to its neighbouring or interacting elements, as well as to both the inputs from and/or outputs to the external environment.

Systems engineers, Kossiakoff similarly argues, "must guide and coordinate the design of each individual element as necessary to assure that the interactions and interfaces between system elements [e.g., subsystems, assemblies] are compatible and mutually supporting. Such coordination is especially important when individual system elements [e.g., subsystems] are designed, tested, and supplied by different organizations [...] Human–machine interfaces and other people-system interactions are particular concerns of systems engineering" (Kossiakoff et al, 2011). Further, as the next lower design level (i.e., assemblies) is defined, adjustments to the parent elements (i.e., subsystems) will inevitably be required, which must in turn be reflected in adjustments to their previously defined interfaces.

In addition to designing "interfaces", the designer ought to similarly appreciate that the "performance of a system relates to the decisions [or the seemingly 'rational' choices] that people make about how to use it" (Simon, 1956; Wallace et al, 2005). In many instances, users may deliberately or otherwise find alternative usages of designed systems. For instance, a warehouse may be utilised as a makeshift dormitory. How will temperature be controlled in cold periods? What about "snoring" pollution and other hygiene needs? What about pilferage among the tenants, In such situations, system performance would be compromised and can even turn injurious.

Nevertheless, situations might arise whereby the design engineer is expected to rather *deliberately* produce a flexible design that accommodates future scenarios, having the ability to undergo change, evolve with relative ease and efficiency—especially when new requirements, needs or possibilities/vulnerabilities emerge over time (Lessard et al, 2014).

NASA encourages such an approach:

> "In some decision situations, the selection of a particular decision alternative will have implications for the long term that are very difficult to model in the present. In such cases, it is useful to structure the problem as a series of linked decisions, with some decisions to be made in the near future and others to be made later, perhaps on the basis of [clearer] information to be obtained in the meantime [...] In these cases, it is desirable to consider

attributes such as 'flexibility' [i.e., the ability of a decision to support more than one current application] and 'extensibility' [i.e., the ability of a decision to be extended to other applications] [...] For example, in choosing an architecture to support lunar exploration, one might consider extensibility to Mars [or perhaps Venus] missions." (Systems Engineering Handbook, 2007)

Project design concepts relate combinations of form to desired functions; each detailed *form—function* combination can be thought of as a design solution (Lessard et al, 2014). In the Large Infrastructure Projects industry, designs ought not to be prepared *directly* on the basis of requirements specifications. The Design Process should be subdivided into a series of *interim* but purposeful stages/steps that would promote manageability, reproducibility (thus, bring about the much-needed consistency), and efficiency of this particular process.

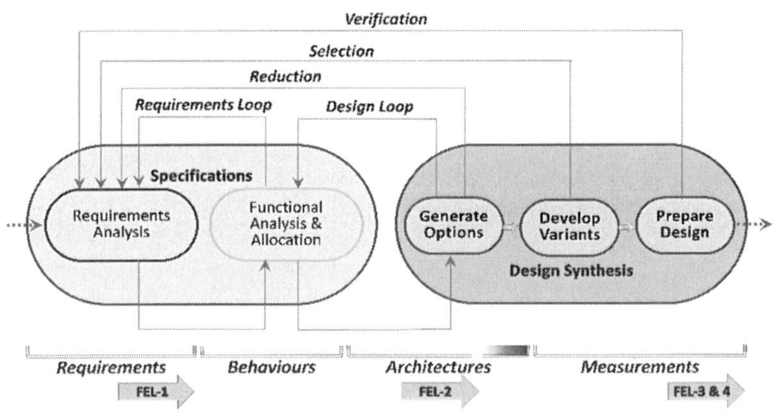

Figure 3.1 – Engineering Design Process (Adapted: ProRail, 2013)

In fact, "The Detailed Engineering design of the system is performed during this phase. Because of the scale of this effort, it is usually punctuated by formal design reviews [to prevent or detect potential errors]. An important function of these reviews is to provide an opportunity for the customer or user to obtain an early view of the product, to monitor its cost and schedule, and to provide valuable feedback [i.e., validation] to the system developer" (Kossiakoff et al, 2011)—another way for engineers to "manage" resources. While essential in theory, the involvement of the customer in the Design Process, in many cases, depending on the "timeliness" or otherwise of their involvements, may prove an inhibiting influence on the design, as many cooks spoil the broth!

This "*spoiling of the broth*", including jumping into Detailed Design, can only be prevented when a structured design process is put in place, be it for the "Specifications Process" (as discussed in Chapter 02) or for the "Design Process". The Dutch MIE recommends *generic* steps for the Detailed Design process applicable to megaprojects:

(1) Generate Options—the central objective of this Option Generation and Reduction process is to determine or identify possible solutions. For instance, a limited number of feasible options for further investigation. Only options that are *inherently* capable of meeting most of the requirements and of achieving the *intended purpose* are transformed into solutions (ProRail, 2013).

(2) Develop Variants—the objective of the Variant Development stage is to be able to make a design choice for the system under consideration that best meets the requirements and other criteria. Further development of the variants is needed to bring the feasible options to a level of detail that allows variants to be mutually compared using a Scoring or Trade-Off Matrix on the basis of specified criteria. Such criteria could include overall costs, ease of production, ease of maintenance, environmental, and prospects and/or liabilities concerning disposal (ProRail, 2013).

(3) Prepare Design—the design selected as the solution of the system (i.e., the most viable variant) is developed in further detail; the requirements affixed to subsystems, subassemblies or the process (i.e., the "How to build") are defined in specific terms at the desired level of detail and all "*forms and dimensions*" specified to cater for the ensuing Construction Process. When the complexity of the design increases (as in large and complex infrastructure projects), not only the "How to build" similarly becomes more complex, but its criticality to the project success would also increase (ProRail, 2013).

The foregoing design approach is not peculiar to the Dutch Ministry (MIE). "In the development of major defence and space systems [and in the development of Large Infrastructure Projects], the engineering design effort [i.e., EDD process] is performed in two steps: [1] designated preliminary design, and [2] detailed design, respectively" (Kossiakoff et al, 2011).

Having investigated the engineering of many such complex systems, Kossiakoff remarks: "Although preliminary design is typically started under systems architecting, many official programs continue to establish a sub-phase where the initial architecture is translated into a preliminary design.

Each step is followed by a formal design review by the customer before the succeeding step is authorized" (Kossiakoff et al, 2011).

He further elaborates on the central purpose of Engineering Design that:

> "The purpose of this highly controlled process [of Engineering Design] is to ensure very thorough preparation prior to commitment to the costly full-scale implementation of the design into hardware and software [i.e., Assembly]. This general methodology, without some of its formality, may be applied to any system development." (Kossiakoff et al, 2011)

Therefore, the ultimate result of the Design and Development Process is a Production-ready Design (e.g., Issued for Construction or IFC drawings) for the system to be constructed/assembled, which may entail many iterations. Ryschkewitsch indeed concurs as follows, "Neither the world's greatest design, poorly implemented, nor a poor design, brilliantly implemented, is worth having" (Ryschkewitsch et al, 2009)—thus, both aspects are required.

In addition to those "Design Steps", Figure 3.1 also indicates significant the information pertaining to "Requirements – Behaviours – Architecture – Measurements" (which were not mentioned in Figure 2.2), proposing as well as at which FEL phase each should apply. Indeed, Scott fittingly articulates these concepts (i.e., Requirements, Behaviours, Architecture and Measurements domains) in "*Engineers behaving badly*" (Scott, 2014), explaining in details the dangers of jumping from "Requirements" to "Architecture" (i.e., behaviours may suffer).

A requirement is a collection of capabilities originating from the owner and/or users (and other stakeholders) that must be met by the solution to the problem and achieve the objectives (Kasser, 2015)—Good requirements at every system-level are critical to the success of any project (Kasser, 2013).

Scott (2014) notes that "requirements" are classified in four levels, namely:

(1) **System Requirements—*what functions shall the overall system do*?** A requirement is a collection of capabilities (i.e., specific set of functions) originating from the users and other stakeholders that must be met by the solution to the problem or achieve the objectives
(2) **Behaviours—*how shall the system interact*?** Functional Behaviours (i.e., interactions between system-elements and with the surrounding environment) allow the synergy (from the combinations of such interactions) that satisfies the needs of the users/customer(s). When "behaviours" are poorly addressed at subsystem level, the overall

system would *behave badly* (i.e., wrong/undesired unanticipated or "emergent" behaviours may impair the system's overall operations)

(3) **Architecture—*what shall the system consist of?*** Architecture is a collection of fundamental concepts and/or properties of a system in its environment embodied in its elements, relationships, and in the principles of its design and potential evolution. Distinct and critical relationships are typified by "interfaces" where data, energy and/or matter are exchanged between subsystems. The complexity of this architecture will also inform the "How to build"

(4) **Measurements—*how well shall the system perform?*** This is the ambit of Detailed Design that determines the size, shape, levels and technical details of parts, as well as performance characteristics of parts, subassemblies, assemblies, subsystems, and overall system

A design engineer (with 25 years of experience) confessed to the author that he has never applied consideration (2); "*I glide from system requirements straight to some architecture*"—he then admitted that his projects did suffer.

To provide a simple yet practical and pertinent example of these concepts, one will use the *wristwatch* (which is actually a "system", as common and immaterial as it might look on the next person's wrist). The relevant *system requirements* considerations are: (1) What shall the *wristwatch* do? (2) How shall the *wristwatch* interact (both internally and with the environment)? (3) What shall the *wristwatch* consist of? And finally, (4) How well shall the *wristwatch* perform?

The following sections will seek to design and develop a new *wristwatch*; and in so doing, they shall provide structured and accurate answers to the aforementioned *systemic* considerations pertaining to the *wristwatch*. Each section will focus on a specific *level* of requirements, leading up to the next:

(1) System Requirements: What shall "come out of and/or go into" the said system? Considering the desired *wristwatch* as a closed "system-box", these "external" components of material, energy, information, and 3-D port are identified as input and/or output (however, to keep it simple, any spatial/temporal constraints are ignored at this stage):

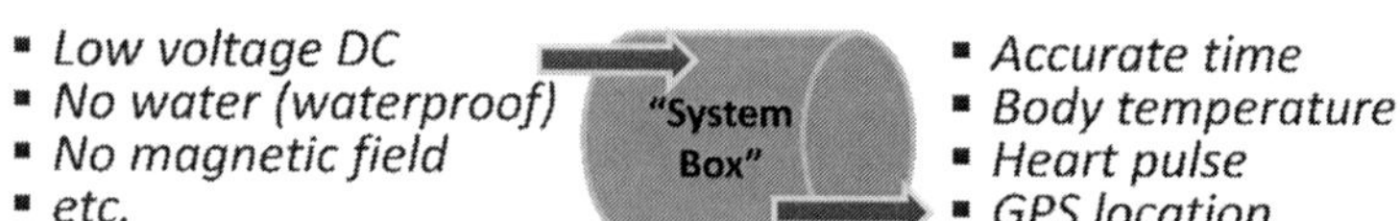

Figure 3.2 – Example of System Requirements (for a 'wristwatch')

Indeed, Figure 3.2 suggests that the new *wristwatch* will transform an input of low voltage electricity (from a 1.5 Volt DC source) and other specified "physiological data" into accurate display of time, body temperature, heart pulse and GPS location respectively—provided that, e.g., NO water ingress is permitted, and that NO magnetic field interferes in its workings.

(2) System Behaviours: How shall the *wristwatch* "function/interact" to produce the intended outputs? The following simple functions and interactions may be identified for the planned product:

Sample of Selected Behaviours for the Design of a New Wristwatch			
01	**Display**/Record/Transmit data	07	'Read' heart pulse
02	Convert DC to 'Eternal timer'	08	Display heart pulse
03	Count time on 'Eternal Timer'	09	'Read' GPS location
04	Display time, on external screen	10	Display GPS location
05	'Read' body temperature	11	Stop/prevent water ingress
06	Display body temperature	12	Shield from magnetic field

Table 3.1 – Example of System Behaviours (for a 'wristwatch')

In particular, Table 3.1 specifies that behaviour #01 could have translated into any suitable combination of displaying, recording and transmission of *wristwatch'* information. For the sake of simplicity, once again, it will be assumed in the next steps of this exercise that the *wristwatch* will only 'Display' the required data (i.e., assuming a different Architecture would be required to 'Record' or 'Transmit' *wristwatch'* data).

(3) System Architecture: What "parts/interfaces" in the *wristwatch* would yield the desired behaviours?

In both theory and practice, System Architecture is not only about "components", but also about any "interfaces" (e.g., interactions and/or interconnections) that shall be in place and in working order to yield the desired System Behaviours. For "*behaviours emerge from interactions between components of a system*"; accordingly, Table 3.2 (see below) provides a sample of selected components.

Architecture for a New Wristwatch with "Display" Function only			
01	Eternal Timer 'chip'	05	Display switch, linked to 04
02	Temperature & Pulse 'readers'	06	Low voltage battery
03	GPS 'locator'	07	Waterproof encasing
04	Multiple Digital Display, MDD	08	'Magnetic Shield' sheet

Table 3.2 – Example of System Architecture (for a 'wristwatch')

Further to the above listing of the components aspects of the proposed Architecture, Figure 3.3 (below) seeks to provide an example of interfaces applicable in this case.

(4) System Measurements: How well (and to what) shall the *wristwatch*'s elements "conform/perform"?

This step, which particularly pertains to the ambit of Detailed Design, is meant to determine the "*shape, size, and technical as well as other performance requirements pertaining to the parts, components and subassemblies of the envisaged system*". Essential to such system measurements is the necessity to verify and validate whether the previously discussed interactions (among system elements) are indeed yielding the right outputs.

Given that Detailed Design is only discussed and elaborated upon in later chapters, will prove premature and, thus, futile to attempt to address the aspects of "shape, size, and performance and tolerance targets" for the *wristwatch* on a detailed basis (i.e., providing dimensions and tolerances) at this stage, except to note that time will be displayed in hours, minutes and seconds, for instance, with an accuracy (tolerance of error) of ± 0.04 seconds during the "Testing" of the system under the EDD process—but might evolve otherwise in the ensuing operations (i.e. utilisation life) depending on its maintenance.

This "systemic and systematic" design approach could as well guide the engineering team in designing a factory or a specialised hospital—by *first* agreeing as to "*What the facility shall do*" (i.e., what/who shall go in and what/who shall come out). Flowing from the understanding of such "requirements", the design team will proceed to decide on the functions (i.e., behaviours) that are expected of the facility in order to fulfil that purpose. The next step would entail allocating such functions to components and interfaces (i.e., architecture)—which will then be dimensioned (i.e., shape, size, performances) through Detailed Design.

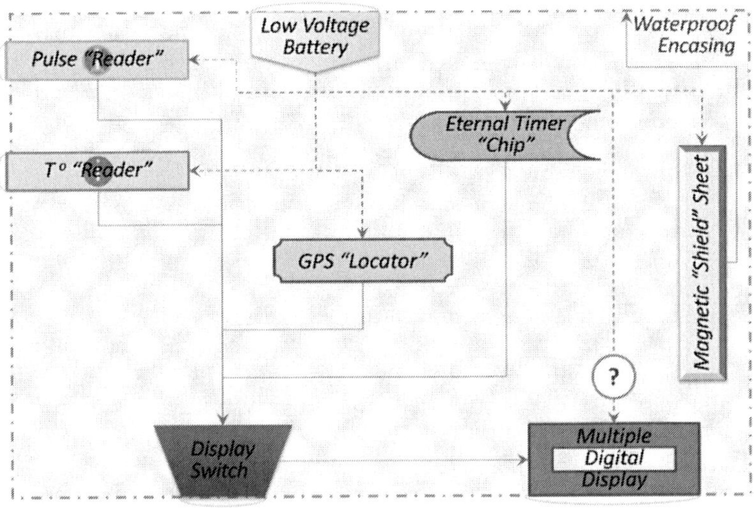

- *A single Multiple Display (linked to a Display Switch) is used rather than self-contained ones*
- *Depending on the final detailed design, the DC supply to Multiple Display might not apply*
- *This "architecture" might bear "emergent behaviours" such as "– ilities" (e.g. reliability).*

Figure 3.3 – Graphical Example of System Architecture (referred to as Embodiment Design)—for the 'wristwatch'

In light of the above, the objective of Detailed Design should be "[...] to produce a complete description of the end items constituting the total system. For large systems, a massive engineering effort is required to produce all the necessary plans, specifications, drawings, and other documentation necessary to justify the decision to begin fabrication" (Kossiakoff et al, 2011). The Detailed Design process, which follows the Specifications process, is intended to prepare the ground for the Construction/Fabrication process. Again, the process must flow from system to subsystems to assemblies to parts and vice-versa in Construction.

Engineering teams would go through a *learning curve* as far as design is concerned.

> "The amount of effort to produce a detailed design of a particular component depends on its 'maturity', that is, its degree of previously proven design. For newly developed components, it is usually necessary to build prototypes and to test them under simulated operating conditions to demonstrate that their engineering design is valid" (Kossiakoff et al, 2011).

The more engineering teams practise Systems Design, the smarter they get at doing it. Nevertheless, the "holistic" system designer will deal with two aspects of Engineering Design and Development (EDD), namely, Technical Leadership and Systems Management as follows:

(i) "[1] Technical Leadership focuses on a system's technical design and technical integrity throughout its life cycle [...] [2] Systems Management focuses on managing the complexity associated with having many technical disciplines, and multiple organizations", affirm Ryschkewitsch et al (2009)

(ii) "[Thus,] once a credible design and its architecture are established, the systems engineer's job is to maintain technical integrity throughout the complex system's very rigorous and challenging life cycle phases" (Ryschkewitsch et al, 2009; Roy, 1993)

"The planning, assessment, and control should not stop after the initial development phases, nor [should] systems analysis, requirements validation, system verification, and end product validation commence well before a physical product is available [...] [they] occur linearly, but with feedback and iterations", it is maintained (Kossiakoff et al, 2011).

> "The process of component design is iterative, just as we have seen the system development process to be. This means that testing must be an integral part of design rather than just a step at the end to make sure it came out properly. This is especially true in the design of components with new functionality, those utilizing unproven implementation approaches." (Kossiakoff et al, 2011)

In discussing "functions-to-components" progression, Kossiakoff remarks:

> "The appropriate process in such cases is "build a little, test a little" [i.e., Verification and Validation (V&V)], providing design feedback at every step of the way. This may not sound very orderly but is often the fastest and most economical procedure. The objective is to validate the large majority of design elements at lower levels [e.g., parts, subassemblies], where the results are more easily determined in less complex test configurations and errors corrected at the earliest time." (Kossiakoff et al, 2011)

Once again, "systemic" feedback loops ought to be in place to validate and/or rectify the iterative transition from Functional Allocation and Design Steps (see Figure 3.1). These "systemic" and systematic feedback loops are discussed in the subsequent sections as follows:

- **Design Loop**: The Design Loop represents the verification of the fit of the design (of the system) and the subsystems with the requirements. It may even become evident during the preparation of the (system) design that a particular allocation of requirements or functions is not desirable. New functions not yet described may emerge, in which instance it is necessary to return to the Functional Analysis and incorporate such insights into the previous Functional Specifications (ProRail, 2013).
- **Verification Loop**: The Verification process is initiated once the Design Process is complete (e.g., testing/inspection of designed components). The objective of the Verification process is to *objectively* demonstrate and to duly document that the specific design matches the requirements; therefore, any material deviations must be noted to initiate corrective actions thereto—and for audit purposes (ProRail, 2013).

For instance, when design specifications for any subsystem or assembly are prepared, a proper V&V Plan shall be prepared for that particular subsystem or assembly as well. Such precautionary measures are required to avoid instances of "*requirements that cannot be verified*" and prevent *pernicious* systems archetypes such as "fixes that fail"—creating solutions that rather address symptoms (i.e., they fail in validation) and, thus, are likely to breed *unintended* consequences (Meadows et al, 1972; Meadows et al, 2009).

This Systems Thinking perspective is surely pertinent to most LIPs, which are complex systems. To be able to properly manage the complexity of a LIP system, a top-down approach to the engineering process is essential, a process repeated more than once from the highest down to the lowest level, based on system complexity (ProRail, 2013).

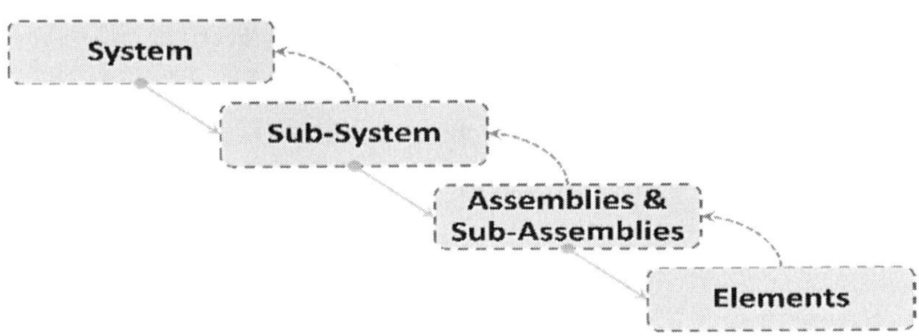

Figure 3.4 – System Hierarchy as a Guide to the Design Process

This "System Hierarchy" reflects a waterfall model. "The waterfall model with feedback [...] depicts the adjustment of inputs from a preceding step to resolve unexpected problems before proceeding to the subsequent step. The waterfall model corresponds most closely to the conventional system life cycle" (Kossiakoff et al, 2011).

In particular when it comes to infrastructure, Systems Approach is the order of the day:

> "As it is being propagated all over the world, companies in general have to deal with fast changing and globalized markets, increasing product and technology complexity, interdisciplinary expertise, continuous innovation and market uncertainty. The approach to the market is interdisciplinary [i.e., based on SE], due to the fact that customers demand more and more integrated [infrastructure] solutions, not only isolated products or services [...] The global behavior of the system [as a Complex Adaptive System, further discussed in Chapter 05] emerges, then, as the effect of the non-linear combination of the interactions among the diverse components [of the infrastructure and with their environment]." (Eduardo et al, 2000)

Further, Scott puts forward the "*Law No 5 of Effective Systems Engineering*" as follows: "To catch [or design] a system you have to think like one" (Scott, 2012). Thus, one shall approach the design of an infrastructure as a system, keeping a "system's view" throughout the design. This will require maintaining the "System Hierarchy" in sequencing the design of its various elements, from systems to subsystems to assemblies to parts that are to be dimensioned (i.e., shape, size, technical performances) via Detailed Design.

At any level of design, engineers shall also consider one level up and down. For instance, when designing a power plant, the power grid and coal supply regime should be considered (for the up or meta-level), while the generators, turbines, and boilers will be considered (for the down-level)—further, those engineers shall explore how such adjacent systems might *evolve* over time.

The purpose of a "system" is found at the level above, its mechanism at the level below. Thus, in designing a coal-fired power plant (as a system), the engineering team ought to carefully consider its *interconnections* with the power grid and the availability and supply of coal. While the purpose of the plant shall be to increase/stabilise electricity generation, this objective may prove strongly dependent upon the quality/quantity of the supplied coal. The fact that the electricity demand/shortage and the coal supply curves may change over time should be considered in discussing the pertinent questions as to "*why, what, for whom, when, where, how*" to build that power plant.

Likewise, the operability/reliability and the technical maturity profile (i.e., new/innovative versus obsolescent products) of the plant's components such as 'generators' and 'boilers' will affect the *overall* performance of the system in terms of Operations and Maintenance support, Mean Time Between Failures (MTBF), availability of spare parts, and disposal dates. The systemic solution(s) provided ought to also accommodate such considerations.

The abovementioned systemic perspectives (i.e., up-system and down-system) are crucial for the "embodiment" of the systemic solution—but are often taken for granted in design. Is it not the "integration" of parts (i.e., 'foundation', 'floor', 'walls', 'roof') that would cause the "emergence" of properties (e.g., *"Providing comfort, shelter against the intruders and the elements"*) for the house being built? The mechanics of the 'foundation', 'floor', 'walls', and 'roof' are integrated to attain the purpose of the planned house; designers ought to perceive their solutions from both perspectives.

Large Infrastructure Projects such as satellites, aircrafts, airports, power or water plants, road/rail networks, mine shafts, hospitals, and shopping malls should always begin with a problem (i.e., perceived at any systemic level) that does, therefore, require a systemic solution. The system to be designed is the "Technical Solution" to the defined technical problem as identified at a particular system level of System Hierarchy as per the below examples:

(i) The Problem X_1 statement consists of the overcrowding problem in the library/study area of a school. The "*system*" to be designed in that case is a "sizeable, high-tech" library/study area to relieve congestion at peak study times (Solution Y_1)—both Problem X_1 and Solution Y_1 are positioned at the same *subsystem* level (i.e., library).

(ii) The Problem X_2 statement is about the clamour in the library during a hailstorm. The *"system"* to be designed here consists of 'noise-cancelling' roofing (Solution Y_2) for the library/study area to prevent clatter during hailstorms—both Problem X_2 and Solution Y_2 are at the *assembly* level (i.e., library roofing) of the school as a system.

In yet another relevant example, the design of an airplane (as a system) will start at the system level by first defining "*What it takes for the airplane to fly in its intended environment*"; then the design focus will shift to its wings (at the subsystem level), and next to the jet-engine (at the assembly level), and further to its blades (at the part-level).

This makes sense on the principle that System Hierarchy shall govern Architecture:

> "The SoS architecture is a living and evolving document that should represent an enterprise view rather than the structured hierarchy of constituent systems. It should set the context for design and implementation of an SoS and establish and document significant decisions about the organization of the SoS and structural elements and interfaces comprising it. The SoS architectural analysis and development is an iterative process of elaborating, partitioning, decomposing, and allocating the SoS functional and performance requirements across the constituent systems. The focal point of an SoS architecting is significant design decisions that are both structurally [e.g., "*less than 7±2 parts*", Miller's rule] and behaviorally important and have a lasting impact on the SoS as well as its constituent systems' performance, evolution, and resilience." (Jamshidi, 2009)

The airplane example aligns itself to the basic definition of a system: "A combination of interacting elements organized to achieve one or more stated purposes" (Haskins and Forsberg, 2011). This top-down approach assures lower-level items contribute to the "purpose". Designing the "airplane" the other way round (i.e., from parts up to the system) will not guarantee that the overall system will achieve its intended purpose, which in this instance is "*To fly in its intended environment*"—do so safely and effectively. Likewise, Table 3.3 (below) provides an example of a system's view of the Technical Solution relating to the design of a "Water Desalination Plant" (by Reverse Osmosis).

This *iterative* engineering process at different levels of detail (of the System Hierarchy) can be linked together by way of Decomposition, Traceability, Design Integration and Requirement Derivation. The next chapter (refer to Chapter 04) shall further elaborate on these notions, as well as provide a depiction of their mutual connections as part of the Design Process.

Hierarchy	Problem – "What"	Solution – "How"	Component
1 System	*Seawater has high salt content and waste, not safe for drinking.*	*Reverse Osmosis (RO) removes minerals from saline solution.*	Desalination Plant using 'RO' method
2 Subsystem	*Impurities and minerals large in size to be removed from solution.*	*Partially permeable membrane to remove ions/molecules, scraps.*	RO Chamber
3 Assembly	*Salt, as the last impurity, has to be removed from the solution.*	*Allow water molecules to pass, but impurities to be flushed away.*	Salt-rejecting Membrane Cast
4 Element	*Separate desalinised water from the brine solution to be exited.*	*Create spiral flow to split water (desalinised) from brine solution.*	Plasticised Tricot

Table 3.3 – Example of System Hierarchy and Problem versus Technical Solution

Just as the up-system (meta-system) and down-systems (i.e., subsystems) were to be considered for the coal-fired power plant, the same perspectives ought to be included in this Desalination Plant example as follows:

(i) The shortage of drinking water and the need to resort to seawater treatment—and any potential change to the initial situation (i.e., a future upsurge in water demand, Authorised usages for *toxic* brine);

(ii) The foreseeable developments in RO technology (and similar) and the technical maturity (i.e., availability, alternative designs) of parts as listed in Table 3.3 as well as any Operations and Maintenance burden of the planned RO Desalination Plant;

Therefore, the design implications to the Desalination Plant include:

(iii) Concerns and insights/foresights arising from (i) and (ii) shall serve as input to the design and development (EDD) process of the Level 1 system (Desalination Plant).

CHAPTER 4

DESIGN AND DEVELOPMENT: THE VEE MODEL

Petroski holds the view that it takes the "*understanding and appreciation of the process and products*" of engineering to access its essence. He remarks: "While the practice of engineering may involve as much technical experience as the poet brings to the blank page, the painter to the empty canvas, or the composer to the silent keyboard, the understanding and appreciation of the process and products of engineering are no less accessible than a poem, a painting, or a piece of music" (Petroski, 1992). To engineer is as *human* as playing music or painting. Engineering, furthermore, is a "pragmatic" matter.

> "The only sure way of knowing if an integrated product is good is to perform verification and validation [V&V] at each stage. The knowledge gained from verifying this integrated phase product can also be used for planning the verification of the final end products." (Systems Engineering Handbook, 2007)

As Figure 4.1 suggests, a process is required that allows for the tracking and verification of design information, from stakeholder identification to problem definition (with input from stakeholders) to translating such into requirements (approved by stakeholders) which are to be converted into a solution that shall solve the (defined) problem, and do so in a way that satisfies the stakeholders.

Figure 4.1 also shows the three domains relevant to SE—the Problem Domain (in which stakeholders define the *Undesirable Situation*), the Solution Domain (in which the new system is created to replace the *Undesirable Situation* with the *Feasible Conceptual Future Desirable Situation* or FCFDS), and the Implementation Domain (in which the new system is being realised) (Kasser, 2015b). (Note that Scott (2012) considers a different perspective in Section 10.3; "Every system design involves three systems"—Law No #9 of effective SE. Scott instead refers to the Realisation-, Solution-, and Context-Systems.)

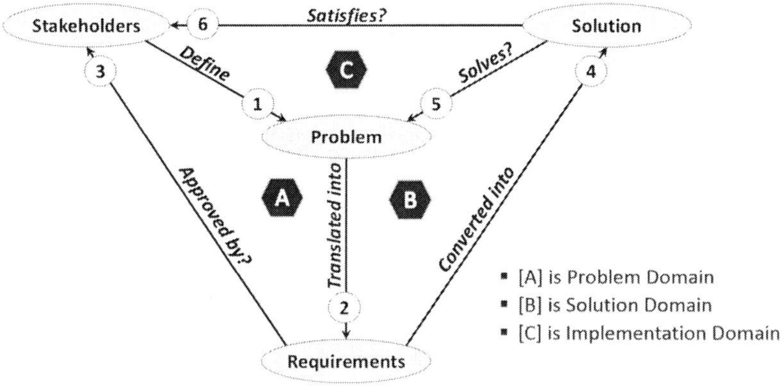

Figure 4.1 – Generic Representation of the Design and Development Process

Kasser explains the difference between these system domains in the following excerpt:

> "It is tempting to assume that the problem domain and the solution domain are the same, but they are not necessarily so. For example, the problem domain may be urban social congestion, while the solution domain may be a form of underground transportation system to relieve that congestion. Lack of problem domain competency may lead to the identification of the wrong problem and lack of solution domain competency may lead to selection of a less than optimal, or even an unachievable, solution system." (Kasser, 2015b)

In a recent infrastructure-related example, the *Undesirable Situation* had arisen from the Covid-19 pandemic in a remote district (i.e., Problem Domain, requiring knowledge in social behaviours and epidemiology). The health response entailed "social distancing" and a specialised hospital (i.e., Solution Domain, requiring expertise in hospital design). Therefore, a derelict abattoir was converted into a field hospital (i.e., Implementation Domain, requiring the wherewithal of a steel construction). All these three domains shall apply.

Thus, any inadequacy of "domain knowledge" will sooner or later lead to risks/difficulties in projects, depending on the project domain in which such a lack of knowledge occurs:

- If in the Problem Domain, the wrong problem may be identified as a result. Solving the *wrong* problem may lead to 'Fixes that fail' syndromes and *wronger* operations

- If in the Solution Domain, it may result in a solution system that would not provide the desired remedy as stated in FCFDS—i.e., sustained improvements in operations
- If in the Implementation Domain, a conceptual solution (i.e., ConOps) may be produced which cannot be realised, such as in designs and execution plans that fail to (adequately) meet the prerequisites and criteria of constructability and operability

The foregoing considerations strongly suggest that any inadequacy or lack of domain knowledge will always compromise the Engineering and Development process (EDD). Yet many project delivery entities insist on *standard processes* over domain competency.

INCOSE, therefore, encourages using the Vee Model "to visualize the system engineering focus, particularly during the Concept and Development Stages" (Fossnes and Forsberg, 2006). The "Vee" highlights the importance of continuous risk (threat and opportunity) assessment, the need to define verification plans during requirements development, and the need for continuous validation with the stakeholders (Fossnes and Forsberg, 2006). Jamshidi (2009) is in support and commends the Vee Model for its *systemic* contributions:

> "The use of such a [Vee] model also significantly improves the traceability required across a product's life cycle by clearly defining functional requirements and system elements."

Such a model guides the development of any new systems and the modelling of their operations. INCOSE, furthermore, maintains that the Vee Model (where time and system maturity both proceed from the left to the right) provides a useful "illustration" of SE activities during the life cycle stages. Accordingly, by applying the Vee Model, project requirements are managed throughout the entire life cycle of the project, from elicitation to their incorporation into the new product (BKCASE Editorial Board, 2011).

The Vee Model caters for a graphical correspondence between, on one hand, requirements and designs, and on the other hand, the realised system-items. This arrangement allows Verification and Validation of "system-elements".

Unless a project delivery entity applies a life cycle model, the "management of projects is still treated in a very amateurish way" (Hamil, 2002; Mabelo, 2016)—one can safely declare. A particular delivery model is increasingly being adopted by the LIPs industry. It stems from the Vee Model and was further adapted by the Dutch MIE (ProRail, 2013) as follows:

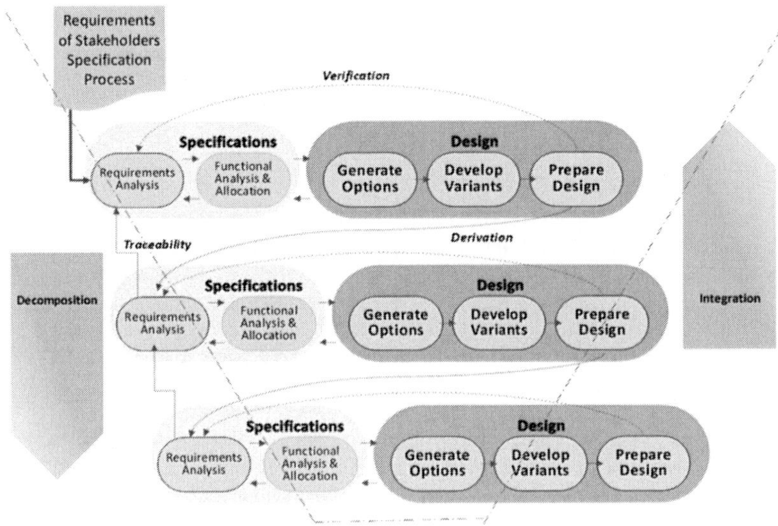

Figure 4.2 – Vee Model Showing the Decomposition Leg (Source: ProRail, 2013)

Moreover, citing Nonaka et al. (1997), Eduardo et al (2000) will maintain that, "There are three characteristics in order for a company [or Project Delivery Organisation] to manage successfully the development of new products process as follows:

(1) To keep a highly adaptive and flexible approach to the development of new products [i.e., new system design], recognising that the development of products rarely occurs in a linear and static way.
(2) To certify the existence of a self-organized [engineering design] team to supervise the development of a new products process. A project team becomes self-organized when it returns to a 'zero information' state, in which the former knowledge is not applicable [to new system design]. Ambiguity and fluctuation characterise this state. The companies need to be willing to give autonomy to the project team and, at the same time, tolerate fluctuations and creative chaos.
(3) To stimulate the participation of non-specialists in the development of new products, which adds variety to the process. This means multi-disciplinarity [i.e., with eclectic project teams], including non-specialists".

The Vee Model, as discussed above, conforms to these requisites and, thus, will prove adequate in guiding and supporting the engineering endeavour related to the design of new systems in the form of LIPs.

Reflecting on the applicability of the Vee Model to LIPs, Bersson observes the following:

> "The Vee model has applicability to complex projects [i.e., Large Infrastructure Projects] that may require [novel] technology insertion, concurrent engineering, and incremental development. The steps flowing down the Vee [i.e., left leg thereof] are User Requirements, Functional Requirements, Configuration and Technical Specification, and Detailed Design. The bottom of the Vee is System Development and/or Configuration." (Bersson et al, 2012)

Wings have recently been added to the "Vee" as part of its adaptation to LIPs to show how project development fits within the broader LIPs project life cycle. The wings are a key addition to the model since it is important to consider the entire life cycle during project development. The left wing shows the regional LIPs Architecture, Feasibility Studies, and Concept Exploration that support initial identification and scoping of a LIP project based on regional needs. The right wing shows the Operations and Maintenance, changes and upgrades, and ultimate retirement/disposal of the system (Ryen, 2008). Engineering Design and Development (EDD) shall always adopt/follow a holistic view. This "*from womb to tomb*" approach is important since, "A system functions [in its *living* environment] exactly as it was 'designed', not necessarily as intended" (Vermaas, 2010).

In keeping with the notion of "System Hierarchy" introduced in Chapter 02 and expanded here, the iterative process of Engineering Design at different levels of detail can be linked together by way of Decomposition, Traceability, Design Integration and Requirement Derivation as illustrated in Figure 4.2 above and further explained in the sections below:

I. **Decomposition:** "The examination of a function to identify sub-functions necessary for the accomplishment of that function, as well as any functional relationships and interfaces […] It consists of repeating the engineering process one level down, subdividing the system into (smaller) subsystems at a lower level of detail, until a level is reached where the design is ready for production" (ProRail, 2013; Fossnes and Forsberg, 2006). For example, the function '*Build Foundations*' could be decomposed into '*Dig Trenches*', '*Place Reinforcements*', '*Pour Concrete*', and '*Allow to Set*' sub-functions.

II. **Derivation:** "The most difficult part of requirements flowdown can be the derivation of new requirements, which often involves a change in the parameters as appropriate to the level in the 'System Hierarchy' [...] The decomposition of a system implies that new functions and requirements are *derived* at a lower level of detail from the choices made during the engineering process at the next higher level of detail [...] depending on the complexity and the system acquisition approach" (ProRail, 2013; Fossnes and Forsberg, 2006). For example, the previous '*Dig Trenches*' sub-function can derive an additional '*Place Solid Barrier*' sub-function for safety purposes.

III. **Traceability:** "A discernible association among two or more logical entities such as requirements, system elements, verifications, or tasks. The requirements at a lower level of detail must collectively produce the requirements at the next higher level of detail, thus linking requirements in order to provide insight into which specific component/subsystem requirements correspond to other subsystems or system requirements." (ProRail, 2013; Systems Engineering Handbook, 2007)

IV. **Integration:** This SE-engine for product realisation processes make up the system structure. It is the first element of the processes that lead from realised products (from a level below) to the realised end products (at the next level above), between the Product/Process Implementation and Verification and Validation (V&V) Processes. There is a need to verify that the subsystems design can be combined to create the total system design, confirming that the designs of all the components at a lower level of detail together form the overall design for the next higher level of detail [...] lower-level products are assembled into higher level products and checked to make sure that the integrated product functions properly. (ProRail, 2013)

V. **Baselines:** Design products such as Specifications can be formally *frozen* as a "baseline" for transition to the next level or for incorporation into contracts […] by ensuring the elaboration of business and technical baselines are acceptable and that they will lead to satisfactory V&V (ProRail, 2013). Thus, "The core of the Vee (i.e., those products that have been placed under configuration control) depicts the evolving baseline from user requirements agreement to identification of a system concept to definition of elements that will comprise the final system" (Fossnes and Forsberg, 2006).

In keeping with the Vee Model, the Engineering Design process (which defines the *design* to be produced) is followed by the Production Process wherein the system is actually and physically produced. Complexity entails that this process also be carried out in layers as well. The said Production Process is, therefore, carried out bottom-up, from the *lowest to highest* level of system detail. Thus, warns Ryschkewitsch, "Neither the world's greatest design, poorly implemented [i.e., defective construction]—nor a poor design [i.e., with errors/flaws], brilliantly implemented—is worth having" (Ryschkewitsch et al, 2009).

Referring to Figure 4.2 above, once the 'Decomposition' leg of the Vee Model has run its full course (i.e., design is complete), the matching leg of 'Integration' will have to start. The Production or Construction process (right leg) consists of a number of sub-processes:

- **Joining or assembling of subsystems**:
  The production-ready design, and the production process requirements (which are of the essence in large and complex projects), provide the key input for this process on the basis of the engineering process. The discrete subassemblies are joined as assemblies and further joined in this process into a set of subsystems that *jointly* compose the system, with interfaces linked to various subsystems to be produced.

- **Verification of subsystems**:
  This process consists of inspection and testing to determine whether a *particular* subsystem behaves "as designed" or "as required" using the criteria/methods described at the time requirements were defined. However, "Many projects have no proper formulation of the metrics, which makes the verification soft or false, or makes the conclusions false", as Professor Clarkson admonishes (Wallace et al, 2005).

- **Validation of the system**:
  This process serves to *objectively* demonstrate that the performance of the realised system meets the needs, expectations, or requirements of stakeholders, after having confirmed that requirements have led to the desired system. Thus, the Validation process may also apply to a "prototype" or "mock-up" of the system under development.

Hood (a requirements expert) suggests the delivery of a system can get stranded between the customer and the contractor when they do not agree on system validation as follows:

> "There are real projects that continued for years, because contractor and customer couldn't reach agreement on the final acceptance, caused by an outdated specification [...] From the V-model [...] it is seen that verification normally begins with the requirements on the lowest and most detailed implementation level [i.e., parts/components] and continues level by level until finally the customer (user) requirements level is reached [...] on this level the system will not be formally verified, but the system will be validated to demonstrate that what has been developed is actually what the customer expects and needs." (Hood, 2008)

In a case study about a *tailored* version of the SE problem-solving process by a team in a complex environment, there was a requirement to solve the problem posed by the need to achieve due consensus on two sets of requirements. Thus, the team proceeded as follows:

> "By considering the cost, priority, and risk attributes of the requirements, as well as clarifying the wording of the requirements for verifiability, both an optimal systems architecture and optimal SDP [System Development Process] process were [satisfactorily] achieved in a relatively short period of time compared to using the standard systems engineering process. Moreover, the customer deemed the Systems Requirements Review (SRR) complete and comprehensive [i.e., following the validation process]." (Kasser, 2015)

Thus, agreeing on the requirements throughout the life cycle is crucial to project success. To illustrate this application to the Vee Model, one will return to the "Water Desalination Plant" project discussed in the previous chapter. By combining the concepts of "System Hierarchy" and "Four Levels of Requirements", the EDD process proceeded as follows:

(A) Decomposition/Design (i.e., Left leg of the Vee Model):

As a "system to be designed", the Plant ought to be 'parsed' into subsystems (e.g., 'RO Chamber', 'Multistage Filter', 'Booster Pump', 'Microfilter', 'High Pressure Pump', 'Energy Recovery Turbine', 'Power Plant' and 'Post Treatment Tank') as per the top-level architecture. Delving into one subsystem, the 'RO Chamber' will in turn be 'broken down' as per the next level architecture into four assemblies (e.g., 'Salt-rejecting Membrane Cast', 'Fibreglass Membrane Shell', 'Brine Spacer' and 'Sealant'). Considering the 'Salt-rejecting Membrane Cast', this particular assembly will be divided into its elements such as 'Plasticised Tricot' and 'Fabric Backing'. Every single "component" of the Plant (i.e., items of the relevant architecture level) shall undergo a Detailed Design process to determine its "*shape, size, performance, and any other technical requirements*" (which refers to the Measurements Domain).

Moreover, this specific part (left leg) of the Vee Model will also entail devising the necessary Construction/Manufacture "processes", Assembly Methods and/or Plans, as well as the relevant Verification and Validation procedures/criteria—as part of the "How to build".

(B) Integration/Production (i.e., Right leg of the Vee Model):

Having completed the Design part (or Left leg of the Vee Model), engineering teams ought to complete the development of the system by proceeding with the Production/Assembly and Integration of the "components" of the system, starting from the lowest level (with discrete elements or parts), then progressing to subassemblies and assemblies.

In line with the specified Construction/Manufacture 'processes', the engineering team shall undertake to 'create/make' and 'assemble' all the relevant elements (e.g., the 'Plasticised Tricot' and 'Fabric Backing') as per the Assembly Methods and V&V Plans to constitute the assembly of the 'Salt-rejecting Membrane Cast'. This item in turn should be joined to the 'Brine Spacer', 'Sealant' and 'Fibreglass Membrane Shell' to constitute the 'RO Chamber' as a "subsystem". Ultimately, all such subsystems will be integrated to deliver the desired "system", being a *ready-to-operate* Water Desalination Plant. If truth be told, this is the ultimate (and indeed the primary) purpose of project delivery; namely, "*to improve or establish operations*". It is not about merely eliciting requirements, nor is it about simply creating facilities.

What this basic example might not have indicated is the set of principles that guided and supported the application of SE processes along the Vee Model. These principles apply to the product-system itself and, therefore, make *irrelevant* whatever specific SE approach is being applied by various systems engineers involved in the project. Indeed, having considered this thorny design issue, Kasser (2015) articulates "*The seven principles for SE*":

> "This Chapter now introduces a set of [seven] updated principles for today's environment [where the needs change rapidly, sometimes even before the solution system is delivered] so that systems engineers working in different domains using various [skills,] tools, techniques and methodologies, can [still] meet the objective of systems engineering [i.e., a useful system]."

The adoption of these seven principles improves SE by suggesting an axiom that applies "to the finished product", the system, irrespective of which SE "approach" is producing the solution system (Kasser, 2015). The set of "axiomatic" principles for the solution system designers are devising should include the following, according to Kasser (2015):

(1) There shall be a clear and singular (set of) objective(s)/goal(s) for a certain system
(2) The ConOps is the first step. There shall be a ConOps from start to finish of system operations, describing the normal and contingency mission and support functions—contingency mission entails the system still *operating* in accident/failure modes
(3) The system shall be designed to remedy the problem over its operational life cycle
(4) The solution system design may be (and often is) partitioned into complementary and interacting subsystems in accordance with the *appropriate* System Hierarchy
(5) Each subsystem is a system in its own right, and shall have its own clear ConOps derived from, and/or compatible with, the ConOps for the whole system of interest
(6) Each subsystem may be (and often is) developed independently, in parallel with other subsystems so long as "*fit, form, function*" and interfaces are duly maintained
(7) Upon successful integration of subsystems, the whole solution system shall be subject to appropriate tests and trials [i.e., V&V of the integrated system], real and simulated, that expose it to extremes of environment and hazards such as might be experienced during operations (as per system ConOps)—thus, allowing for validation.

Furthermore, when designing the solution system, two key tenets should be borne in mind:

(I) "Coping with changes" is a design criterion (i.e., the solution system creation and the realisation process shall be able to cope with changes in needs or requirements).
(II) Costs and Schedules are NOT (nor construed as) design criteria (i.e., these considerations may be used as selection criteria for choosing the single/desired solution system option from among many ConOps, after the solution system options have been designed).

Finally, Hitchins (2009) draws the attention of system designers to yet another principle. "The essence of Systems Engineering is in choosing the right parts [e.g., subsystems], bringing them together in the right way, causing them to interact in the right way, and in orchestrating those interactions to create a unified whole [e.g., intended system] that performs with optimum effectiveness in its operational environment, so solving [and/or remedying] the problem [i.e., the *Undesirable Situation(s)*] that prompted its creation."

Thus, the Desalination Plant engineers should have applied these principles as follows:

(1) The *singular objective* is to devise a Desalination Plant that will treat Z-amount (litres per day) of seawater solution (i.e., containing high salt content and waste)
(2) The overall Concept of Operations (ConOps) consists of, "*Pumping seawater into a tank before boost-pumping it through a series of multistage and micro filters into the Reverse Osmosis Chamber to a Post-Treatment Tank for storage and distribution*"
(3) When in operation, the Plant shall render seawater drinkable by way of removing salt, impurities, and large-in-size minerals from the solution, while exiting brine
(4) Using Reverse Osmosis (RO), the Plant ought to include subsystems as follows: 'RO Chamber', 'Multistage Filter', 'High Pressure Pump', 'Energy Recovery Turbine', 'Booster Pump', 'Microfilter', 'Power Plant', and 'Post Treatment Tank'
(5) Each subsystem must function as per its individual ConOps; e.g., 'RO Chamber' will "*force seawater through a partially permeable membrane*" (i.e., 'Salt-rejecting Membrane Cast', 'Fibreglass Shell', 'Sealant', 'Brine Spacer') to sift salt molecules and scraps
(6) Subsystems could be acquired (e.g., manufactured, purchased as COT) as *separate but complementary, interacting* system-items
(7) A suitable Construction/Manufacture process shall be followed by the engineering team to "create/make", and then "integrate" and "test" all the elements of the Plant

Further to the above axiomatic principles, the following should also *incidentally* apply:

(8) The "design process" of the Desalination Plant would accommodate changes to the required performance (i.e., as per salt content) and to the work methodology (e.g., based on ground conditions that might differ from prior data and surveys)

(9) Cost and (delivery) schedule should not be considered as "design" but rather as "selection" criteria—e.g., to select parts based on COTS availability and/or price

(10) The engineering team ought to make sure that the *right* parts for the Desalination Plant were selected and *adequately* integrated to assure optimum performance of the plant. Thus, a suitable "orchestration of interactions" among the subsystems is crucial

Figure 4.3 below shows the requirements and architectures (i.e., conceptual, functional, physical, integration and interface, operational) pertinent to the Vee Model. Figure 4.4 captures a depiction of System Integration on the right-hand side; the ladder exhibits the bottom-up approach in line with "System Hierarchy". Indeed, Figure 4.4 shows how parts/elements are joined together to form assemblies, which in turn will be joined into subsystems—before those latter are ultimately integrated into the overall "intended" system.

This Construction process (which must not precede, overlap with Design Engineering) is carried out *bottom-up*; namely, from the lowest elements (i.e., parts) to the highest subsystems/system in accordance with the relevant Design and Construction processes.

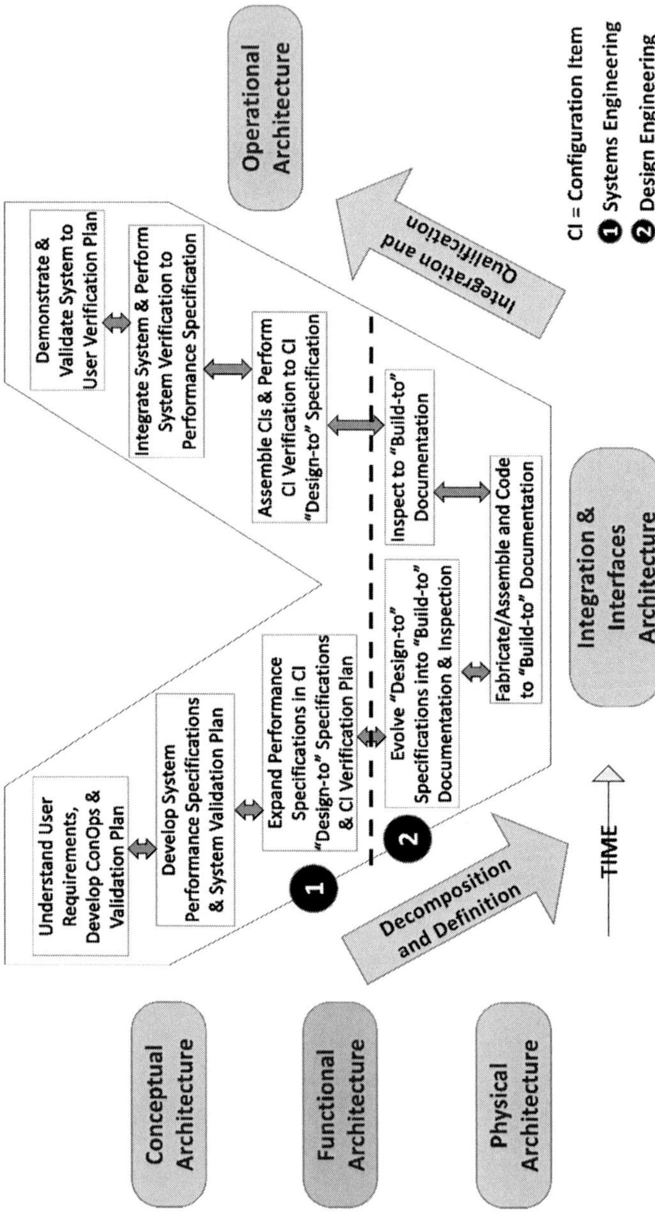

Figure 4.3 – Vee Model Showing Decomposition and Integration (Adapted: Habayeb, 2005)

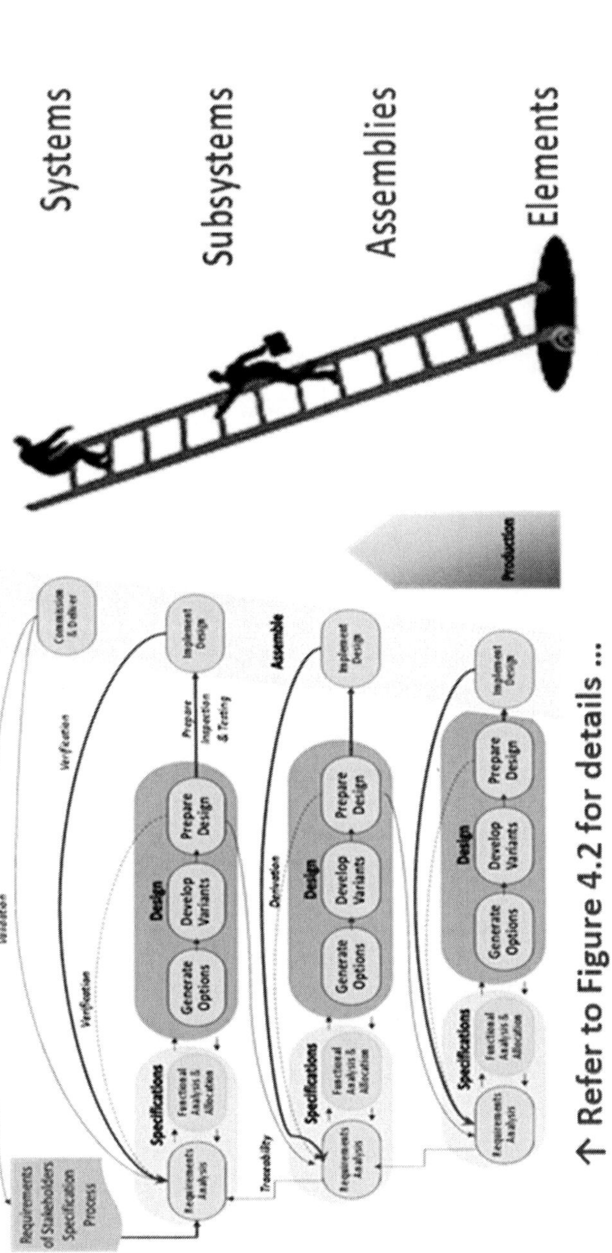

Figure 4.4 – Vee Model Showing System Hierarchy (Adapted: ProRail, 2013)

Any overlapping (also known as coupling) of Engineering Design and Construction must be avoided, particularly in LIPs, otherwise overruns far exceeding 25 % will arise on schedules and costs—leading to complications and project failure (Merrow, 2011). In some cases, however, this coupling may be accommodated at "system level" so long as it is not allowed at "work package" level. For instance, a project's "highlight report" once read:

> "A managing contractor approach was selected, supplemented with an early contractor involvement […] [and] component comprising of multiple [i.e., not coupled] packages released to allow for the design to be progressed while construction proceeded. This form of contract could respond to the time constraint and scope definition." (Australian Government, 2012)

Nonetheless, a large project proceeding on this basis will still involve *higher* delivery risks:

> "The systems engineering method divides the design task into two closely coupled steps: (1) analyzing and formulating the functional design of the system […] and (2) selecting the most advantageous implementation […] The close coupling between these steps results from their mutual interdependence, which requires both visualization of the implementation step […] and iteration of the implementation [...] [thus, breeding difficulties]." (Kossiakoff et al, 2011)

Industry experts are adamant that starting construction without a completed design for the "package" is always a bad idea (Merani, 2020). "Let's start building and we can figure out the design as we go along", bemoans Merari of Projects News. "You don't have to be a construction expert to know that this doesn't sound like a great approach, but this scenario is a lot more common that you'd think, industry professionals in the United Arab Emirates have warned". This rush to deliver projects often means that contractors begin laying bricks "*before ink is dry on the drawing board*", meaning it is no surprise that design discrepancies remain a key cause of disputes and delays on some projects (Merani, 2020).

The same article further reported the following disconcerting statements on projects "*being under development and implementation at the same time*" (i.e., strong coupling):

- Engineers are probably aware of the danger of coupling. "But because with most of the clients, their design is not completed and they develop the design during the construction time, it causes some delays and additional costs for the contractor, and variations and extension of time." (Merani, 2020)

- "Project delays do occur when a client [who is eager to finish] goes to the site without sufficient time for the design to be completed and coordinated correctly." (Ibidem)
- "We are, however, from time to time required to prepare a design and go to the site in less time than we would ideally like to carry out the full [design] coordination. When this happens, it can inevitably lead to a greater number of design discrepancies than the norm [which will then result in overruns]." (Ibidem)
- "When the focus is on trying to save cash, it only stores up problems for a later date; all projects evolve to some extent along the way, however, where a design has not been properly completed and coordinated there is a greater risk of variations [such as in cost] and delays to the programme." (Ibidem)
- "Where, however, sufficient time is given to the [systems engineering] consultant, we don't generally see the same issues. The [reasonable] value in mutually agreeing a realistic programme with achievable deadlines at the design stage can save significant time and additional cost during the [ensuing] construction stage." (Ibidem)

In a nutshell, "*commencing construction before design is completed*" is definitely a recipe for disaster. However, should the design team fail to establish (i.e., in the first place) and adhere to a realistic "design timetable", worse predicaments might still befall the projects.

In addition, the planning of construction works might entail some kind of re-iteration of the initial design in some cases to accommodate additional works for the transition stage. Interim arrangements might also be required:

> "Some plans may have to consider additional problems due to constraints imposed by the infrastructure during the construction and transition phases [...] if additional roads or railway lines have to be constructed, the effect on traffic in the existing infrastructure in those locations during the construction phase will have to be estimated and taken into account." (Kasser, 2015)

Should this consideration not be addressed, construction could be adversely affected (as the existing infrastructure capacity may fail to carry the extra traffic during construction). The author once got to commission a traffic study to determine the specific *easy* slots (i.e., outside of peak times) when construction-induced traffic (i.e., delivery or rubble trucks, abnormal loads) could be accommodated on the already congested roads connecting the highway to the construction site without jamming the entire road network, until project completion.

Kasser (2015) discusses another scenario where an attempt to remedy such a predicament caused more harm than good. A systemic approach should have been applied in this case:

> "In 2011 the company constructing a new apartment building needed to protect the entranceway into the apartment development from small objects that could potentially fall from the upper floors during the construction. So they engineered a solution by constructing a covered area and diverted pedestrians around it.
>
> The pedestrians, when faced with going three quarters [of the way] around a square or taking the short route tended to take the short cut [...] which meant that the pedestrians had to step out into the road: a hazardous situation. The engineers, noticing this undesirable situation extended the fence along the driveway by about 30 metres to stop them from stepping out into the road. This meant that pedestrians had to remember to walk out into the driveway 30 metres before they reached the detour sign.
>
> The new undesirable outcome was an extension of the hazardous situation rather than eliminating it. This was because pedestrians coming the other way walked in the road for an extra 30 metres sharing it with incoming traffic. A little bit of holistic thinking [i.e., Systems Thinking] would have predicted the pedestrian behaviour and might have provided a different safer solution." (Kasser, 2015)

Factoring in human behaviours into a design (i.e., "How to build") is mostly useful. Thus, effective communications with operators and users are essential throughout the lifecycle.

> "Effective technical communications are difficult to maintain in any organization for a variety of reasons, many of them inherent in human behavior. They are, nevertheless, absolutely vital to the ultimate success of the development project." (Kossiakoff et al, 2011)
>
> "From another perspective, most systems cannot operate without the active participation of human operators exercising decision and control functions. In a functional sense, the operators may well be considered to be integral parts of the system. However, to the systems engineer, the operators [or users] constitute elements of the system environment and impose interface requirements that the system must be engineered to accommodate. Accordingly, in our definition, the [community] will be considered to be external to the system." (Kossiakoff et al, 2011)

CHAPTER 5

ENGINEERING DESIGN METHODS

The infamous Tacoma Narrows Bridge (USA) was a scale-up of an old design. But the strait where they replicated it had stronger winds; the bridge became unstable in those crosswinds and it collapsed. Since the design engineers reused the requirements for an existing bridge, it is safe to assume that these requirements were up to the standards of the day. The bridge was also built well, but it was the wrong bridge for that environment such as crosswinds (Bahill and Madni, 2017). The context of the solution does matter.

The cause of the bridge's failure, as scientists later discovered, was "aeroelastic flutter", not resonance (Billah and Scanlan, 1991). However, a contributing factor was its solid sides, not allowing wind to pass through the bridge's deck. Its design allowed the bridge to catch the wind and sway, which ultimately took it down. Flutter is a dynamic instability of an elastic structure in a fluid flow, caused by positive feedback between the body's deflection and the force exerted by the fluid flow (e.g., gale-force wind).

The Tacoma Narrows Bridge was not designed as a Complex Adaptive System (CAS) and, thus, it lacked the "resilience" to survive the adverse interactions with the environment; the bridge lost its stability, eventually failed, and collapsed. So many other large and complex infrastructures might have suffered this fate without engineers realising it or being inspired to adopt a CAS design approach—how many more systems shall still fail?

A CAS is a "dynamic system" able to *adapt to* and *evolve with* a changing (or rather evolving) environment. "It is important to realize that there is no separation between a system and its environment in the idea that a system always *adapts to* a changing environment. Rather, the concept to be examined is that of a system *closely linked with* all other related systems making up an ecosystem" (Chan, 2001). One needs to consider the evolution of adjacent systems so that the solution system being implemented will interface with the adjacent systems as they will in the future at the time of interface, and as they will be after being upgraded, which may differ from their current state (Kasser, 2015). For instance, the design of a coal-fired power plant ought to align to the "life cycle" of its coal mines. Change needs to be seen in terms of *co-evolution with* all other related systems, rather than as *adaptation to* a separate and distinct environment (Chan, 2001).

Even so, the CAS concept shall suitably apply to the design problem at hand. "The definition for CAS seems to change with the different attempts at application. In order to make a good match between a hard-to-solve problem and a complexity approach, it is important to consider whether and how the problem [i.e., design of an infrastructure] exhibits attributes of a complex adaptive system [...] CAS have a number of characteristics" (Chan, 2001).

Research indeed indicates that CAS have a number of characteristics which are described in the subsequent subsections. Infrastructure systems (LIPs) are usually dynamic systems characterised by a large number of interacting parts/components, and complex relationships between these latter, that have to respond and adapt to changes in the technology, and the socio-economic and political environment (Lo Sorto, 2020). In line with this introduction to Complex Adaptive Systems (since a broad treatise on the subject is beyond the scope of this manual), key features of a CAS as described by the Massachusetts Institute of Technology (Chan, 2001) are listed as follows:

I. **Distributed Control**: "There is no single centralised control mechanism that governs the system's behaviour" (Chan, 2001). Although the interrelationships between the elements of the system produce coherence, its overall behaviour usually cannot be explained merely as the sum of [its] individual parts or "elements" (Gould et al, 1996)

II. **Connectivity:** "Complexity results from the interrelationship and interconnectivity of the elements within a system and between a system and its environment. A decision or action by one part within a system will influence all other related parts, but not in any uniform or linear manner" (Chan, 2001)

III. **Co-evolution:** "With co-evolution, elements in a system can change based on their interactions [internally] with one another and/or with their adjacent environment. Additionally, in some cases, patterns of behaviour can change over time" (Chan, 2001)

IV. **Sensitive Dependence on Initial Conditions:** "CAS are sensitive due to their dependence on initial conditions. Changes in the input characteristics or rules are not correlated in a linear fashion with outcomes. Small changes can have a surprisingly profound impact on overall behaviour, or vice-versa. A huge upset to the system may not affect it markedly. Long-term prediction [of behaviours] and control are therefore believed to not be possible in complex systems" (Chan, 2001)

V. **Emergent Order:** "Complexity in CAS refers to the potential for emergent behaviour in complex and unpredictable phenomena [...] each is a system with a network of many agents acting in parallel [...] in each system, each agent is in an environment produced by its interactions with the other agents in the system. There is constant action and reaction to what other agents are doing, thus nothing in the environment is essentially fixed" (Chan, 2001)—this point is relevant in engaging a network of stakeholders. Self-organizing (i.e., adaptive) systems that reorganize as needed are better able to operate more effectively in open systems as compared to controlled systems

"From the interaction of the agents arises some kind of global property or pattern which could not have been predicted from simply understanding each particular agent [...] Any coherent behaviour in a system would arise from competition and/or cooperation among the agents themselves. Therefore, self-organisation is an inherent property in every CAS" (Chan, 2001). For instance, the aeroelastic flutter of the Tacoma Narrows Bridge was an *emergent* behaviour that the system was to adapt to

VI. **Far from Equilibrium:** "When pushed away from equilibrium, a physical system could survive and thrive; should it remain at equilibrium, it would die. The 'far from equilibrium' phenomenon illustrates how systems that are forced to explore their space of possibilities will create different structures and new patterns of relationships [...] Thus, it can be said that CAS function best when they combine order and chaos in appropriate measures" (Chan, 2001)—structures shall be designed to survive chaos

VII. **State of Paradox:** Recent research about CAS has indicated "dynamics" combining both order and chaos. "This reinforces the idea of bounded instability or the edge of chaos that is characterised by a state of paradox [that is dynamic and should be consider in design]—stability and instability, competition and cooperation, order and disorder" (Chan, 2001)

It is globally acknowledged that ICT and large infrastructure systems have distributed controls (i.e., local vs. central units); interact with their adjacent environments; exhibit non-linear behaviours (i.e., with small changes causing major impacts); entail emergent behaviours (i.e., unpredictable outcomes); and allow chaos and/or homeosis. Accordingly, "The design of large IT [as well as of construction] infrastructures is a complex problem because it is dependent on many variables and must evolve rapidly to cope with business changes" (Da Silva and Abreu, 2010). Large infrastructure systems (LIPs) are largely construed as CAS.

Large and complex infrastructures should, therefore, be designed assuming the above CAS features. The design and planning necessary to delimit the project scope and "boundaries" and define infrastructure asset features are often an uneasy iterative process in which many actors (i.e., stakeholders) are involved. For instance, users, engineers, planning authorities, local and national governments, ministries, financial/banking institutions, suppliers, main contractors, and subcontractors develop a shared mental model or cognitive map of what the project should be and how it should be delivered. This long step of a megaproject's life cycle is typified by a large amount of effort and risk (Lo Sorto, 2020; Youker, 1998).

Further, recent developments in the realm of large and complex typified infrastructure have given a new meaning to the concept of SoS. "SoS is clearly about the design and engineering problems of combining existing systems into a larger whole that yields new functionality, not available through any of the constituent systems. These problems comprise the architectural principles of such combination, the engineering process and the technical solutions", Lukkien (2015) submits. Hence, SoS are in essence CAS as well.

INCOSE maintains that, "System of systems applies to a system-of-interest whose system elements are themselves systems [i.e., CAS]; typically, these entail large-scale inter-disciplinary problems with multiple, heterogeneous, distributed systems" (Fossnes and Forsberg, 2006). Jamshidi (2009) states that, "We are, instead, interested in what makes a SoS difficult to deal with in terms of its behavior, from the standpoint of engineering solutions for them. SoS have the following properties [which are found in most LIPs]:

(1) Operational independence of the individual systems
(2) Managerial independence of the individual systems
(3) Geographic distribution (i.e., very pertinent to most infrastructures)
(4) Emergent behaviour (e.g., traffic congestion, sprawling of cityscape)
(5) Evolutionary and/or adaptive development, especially with *morphic* material"

The SAND Report also discusses this remarkable concept of a Complex Adaptive System of Systems or CASoS, in short:

> "A CASoS is a complex adaptive system of systems. Our current definition is cast in terms of the definitions of the component terms. Complex refers to complex behavior, encompassing behavior that is greater than a sum of the parts, exceeds expectations, difficult to understand, [non-linear,] emergent, results from interactions of components" (Sandia Laboratories, 2011).

Adaptive refers here to the ability of the system to change itself in response to external stimuli, whereas "SoS" refers to the structure of the system whose elements are themselves systems. It would, therefore, follow that large and complex infrastructures such as LIPs are basically Complex Adaptive System of Systems (CASoS).

Many design engineers will find quite foreign the foregoing discourse, as it breaks away from their traditional approaches of designing infrastructure as static or "closed" systems, with little to no *dynamic* interactions with the environment. But before someone labels them "*old school*", it must be remembered that the 146.6 m tall Great Pyramid of Khufu in Egypt (like the many more, and perhaps more ancient pyramids in Sudan) was probably designed as a closed system (literally and functionally) and it is still standing after more than 4,600 years (Stille, 2015).

Nevertheless, history tells us that the amazing Leaning Tower of Pisa (56.67 m tall), though still standing owing somewhat to its "lead counterweights", is no longer performing according to its original design of a straight, freestanding bell tower (Shrady, 2003). The tower began to lean during construction in the 12th century, due to soft ground which could not properly support the structure's weight (14,500 metric tons), and it worsened through the completion of construction in the 14th century, tilting some 5.5 degrees. What might have caused it to *behave badly*? The foregoing discussions about CASoS might have the answer.

The unintended and unanticipated "interactions" between the structure of the tower and its adjacent foundations resulted in an uneven settlement under the weight of the marble and stones. Not even the remedial works carried out between 1993 and 2001 could succeed in reversing much of the unwelcome tilt. The original sin was that its designers did not anticipate the "complexity" of the ground settlement to the point of adapting the design of its foundations accordingly. The Tower of Pisa failed to adapt, evolve with its environment.

Hence, unless a large and complex infrastructure is construed and designed as a CAS or CASoS, it might in the end suffer the same, if not worse fate than that of the Tower of Pisa. How else would engineers have designed the 604 m tall Canton Tower in China? The once tallest tower in the world (as of 2010) boasts of a "180 m open-air skywalk" where visitors can physically climb the tower. There are also outdoor (i.e., open) gardens set within the structure, and at the top, just above 450 m, a large "open-air observation deck", as well as the "two-story *rotating* restaurant" (Canton Tower, 2010). However, its real ingenuity rests in the "Tuned Mass Damper" that reduces the amplitude of vibrations (and motion sickness) caused by "environmental forces" acting on a complex structure, such as winds or earthquakes, or by a seemingly innocuous vibrations source causing resonance (i.e., emergence). Such hostile actions could have been destructive, unpleasant or inconvenient.

Approaching the engineering design of major and complex infrastructures from a Complex Adaptive System (either as CAS or CASoS) perspective definitely provides design engineers with a welcome opportunity to influence desirable "emergent behaviours" that could arise from judiciously orchestrated interactions between a system's components. And exploiting any *desirable* emergence starts at the level of "components"—and their various, directed "interactions".

For instance, the author was once managing a project to "upgrade" an inland container terminal in Southern Africa. The project objectives encompassed the modernisation of the inland terminal to handle 0.80 million TEUs (i.e., Twenty-foot container Equivalent Unit) per annum from a current 0.24 million TEUs. The scope involved rehabilitating the infrastructure and refurbishing existing mobile equipment, and the reconfiguration of the terminal to maximise operations.

The "architecture" of this specific terminal included the following systems:

a. Administration building
b. Truck-in and truck-out gates
c. Weight-bridges
d. Rail access line
e. Rail terminal and sidings
f. Stacking yard
g. Mobile handling equipment (i.e., reach-stacker, rubber-tyres gantry)
h. Workshops and technical amenities (e.g., effluent treatment plant).

In addition to the endless list of items in a state of disrepair—such as the "failed" concrete pavement, the outdated and blocked drains, the flawed effluent containment system, the badly damaged rails, and poor security —the terminal was "cramped" in its current confines with no room for expansion. A difficult question then arose: "*How to increase throughput from a current 0.24 mil TEUs per annum (by ± 300%) to the required 0.80 mil TEUs without having to expand the "Stacking Yard" at the terminal?*"

Of course, every other terminal element had to be repaired or upgraded, thus making stronger each component of the terminal; even so, the answer came from appointing a "Simulation Consultant". What really worked was a judicious combination of "Mobile Handling Equipment" and the enhanced terminal-layout that improved both the interconnections and interactions between those elements, causing the 0.8 mil TEUs throughput to "emerge".

Large and complex infrastructures are *first and foremost* CAS. In those increasingly common instances where the infrastructure will consist of "independently useful systems incorporated into a larger system that delivers 'unique' capabilities" (Mabelo, 2016), they shall be deemed and treated as CASoS. Therefore, the concept of emergence shall prevail. The next six sections discuss design methods that treat infrastructure as such, for just as Winston Churchill once said, "*Give us the tools and we will finish the job [i.e., the Engineering Design and Development (EDD) job]*".

> "Effective workmen sharpen their tools. Effective systems engineers [and other project practitioners involved in LIPs] not only sharpen their tools they are also always on the lookout of new [SE] tools that they can adopt or modify for their own use." (Kasser, 2015b)

Appropriate tools are required in dealing with systems with a strong human interaction. "It is generally recognized that societal [i.e., human] systems tend to be more complex than technological systems and this in turn has led to new ideas regarding the appropriate tools to be used for analyzing such systems" (Gottinger, 2012). Gottinger (2012) further argues, "Our analysis shows—roughly—that the global behavior of complex dynamic societal systems embedded in the environment tends to a breakdown (catastrophe, chaos) after finitely many 'real-time computations' [i.e., using complex tools] either because of changes in the environment or because of the 'evolution-complexity' character of the system which, if not sufficiently well understood, may lead to behavioral rules incompatible with [its] survival."

Therefore, design engineers ought to select an appropriate "combination of tools" (i.e., tools that fit the complexity of their system) from the below list.

5.1 The Quality Function Deployment (QFD)

Product/system development (including infrastructure), Shrivastava (2015) submits, is not only becoming more and more complex (Maqsood et al, 2009), but it is increasingly characterised by extended development time (e.g., caused by excessive redesign, problem-solving, failed transition to operations), lack of an efficient and effective product/process development team, lack of structure to the allocation of product development resources, and customers complaining or not being satisfied with the product/system or service. This is even more acute in Large Infrastructure Projects (LIPs).

When planning a new "product/system", engineers would typically examine the manufacturing and performance history of the current product (when applicable); they will also examine customer satisfaction information that might happen to be available. Unfortunately, much of this information is often incomplete. Hence, a design tool is needed that supports information processing and decision-making in the Engineering Design process.

The QFD is a design method that makes use of focused discussions in *eclectic* groups to systematically address product/process and production quality. It uses a matrix format to capture a number of issues that are crucial to the planning process.

QFD was first practised in the 1970's by an engineer at a shipyard in Kobe (Japan). In our time, this method is being used in various fields like automobile, electronics, home appliances, apartment layout planning and clothing—but only sparingly in the infrastructure space. The method allows a design team to achieve the following:

I. Identify (and document) the customer needs regarding a product and their relative importance (i.e., the relative rankings of such needs)
II. Map out the customer needs into specific engineering requirements in the product (so that requirements will best satisfy those needs)
III. Identify the design components or modules (e.g., subsystems) that must be added, changed or modified for each identified need
IV. Prioritise the design developments and/or modifications that must be made to maximise the customer benefit at a given (lowest) cost.

The House of Quality (HoQ) matrix is the most recognised and widely used form of the QFD method. It is used as a tool integrated in larger design processes and to support product/service and process design. In essence, it translates "elicited" customer requirements, based on marketing research and benchmarking data, into an appropriate number of engineering targets to be met by the new system design. This generally portrays the HoQ as "a kind of conceptual map that provides the means for interfunctional planning and communication", according to Hauser and Clausing (1988).

The initial steps in forming the HoQ include determining, clarifying, and specifying the customers' needs. These steps are the foundation for a clearly defined venture and will ensure a product or process is well thought out prior to any further development (Tapke et al, 2001). Therefore, a clear and all-inclusive list of functions and sub-functions focuses on the important needs of the customer; customer wording or terminology is frequently used to express the "voice of the customer". Shrivastava (2015) upholds the understanding of the customers' needs:

> "In the early days, by adopting [a] product-driven approach, [many] manufacturing industries or businesses were introducing their goods into [the] marketplace without considering customer views and needs. But in successful product development, understanding of the customers' needs and requirements is regarded as a key issue."

The strength of HoQ as a methodology is that it gets the right people together (i.e., in eclectic teams), works efficiently and effectively to meet customers' needs and facilitates better decision-making. Thus, it is usually performed by a multidisciplinary team representing sales/marketing, design engineering, manufacturing engineering, and any other functions deemed critical (in "value" creation) by the company (Shrivastava, 2015).

The HoQ matrix may take different forms. Assuming designers have come up with several possible improvements and/or variations, this methodology will assist in determining which ones to incorporate. However, its ability to be adapted to the requirements of a particular problem makes it a very strong and reliable system to use in varied circumstances; this supports using it to devise complex infrastructure systems (e.g., *integrated* transport system).

The "House of Quality for Product Planning" matrix, which is actually a very popular design tool and the "starting step" of the QFD, systematically structures the design information as follows:

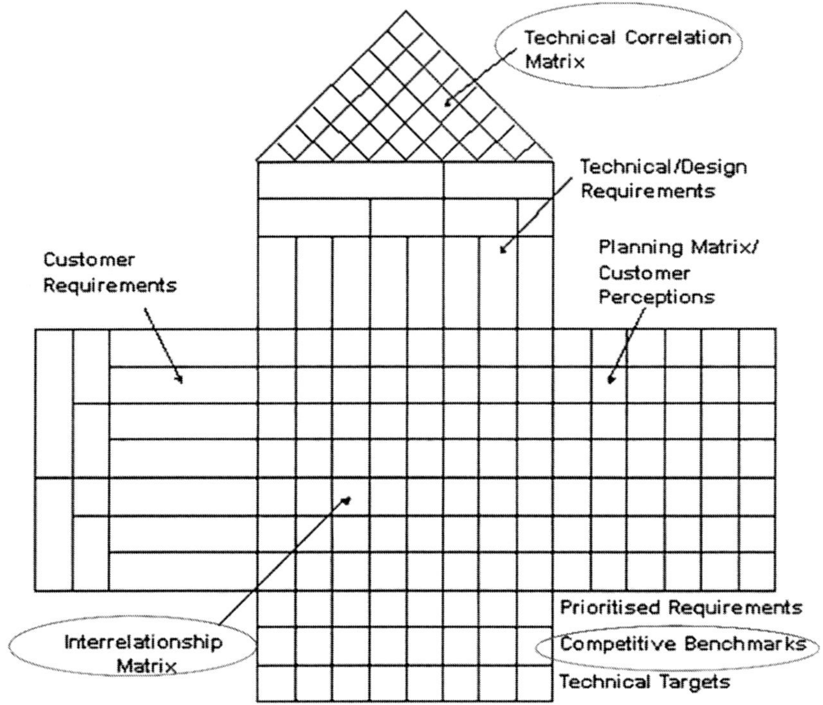

Figure 5.1 – House of Quality as per a Generic Model (Adapted: Tapke et al, 2001)

i. Customer Requirements:

These requirements are stated as customer/user characteristics, e.g., faster, smaller, cheaper, greener, greater esthetical appeal, stealth or soundproof.

ii. Customer Importance Weights:

These serve to rank or prioritise the customer demands, based on statistical analysis of surveys and other data collection techniques.

iii. Engineering Characteristics:

These translate customer needs into clear/concise technical specifications, i.e., measurable engineering requirements or "technical parameters".

iv. Correlation Ratings of Requirements and Characteristics:

These complete the relationship matrix; the goal is to associate a technical feature to each need of the customers which it affects.

v. Benchmark Satisfaction Ratings:

These result from analysing competing products, which are ranked in terms of both customer satisfaction and technical performance.

vi. Relative Technical Difficulty:

This relative ranking of the technical difficulty of achieving each technical specification is set based on organisational capabilities.

vii. Technical Correlation:

Having identified the pertinent inter-relations between different technical parameters, the "roof" is completed to indicate that achieving a given design requirement might involve some conflict resolution.

viii. Engineering Specification Target Values:

These are based on the customer requirements and the values of the benchmarked products. The goal is to prioritise the exercise by improving the most critical design issues first, given limited ability.

Customer satisfaction is a crucial goal to be accomplished at an affordable cost. The HoQ Matrix is developed to satisfy most customer requirements (Shrivastava, 2015). "Customers buy benefits [i.e., which satisfy their needs] and producers offer features [i.e., engineering characteristics] [...] It is of utter importance to translate the wishes [e.g., parts, costs, functions, quality, reliability] of each and every customer into some tangible values that can be turned into engineering specifications [...] Each aspect of the desired item must be clearly defined: measurements must be defined, heights specified, torques stated, and weights targeted [...] In any event, the needs, which were clarified and then explicitly stated, should be satisfied to the best of that organization's ability" (Tapke et al, 2001).

A practical example of the application of HoQ tools is provided in Section 12.11. Nevertheless, while gaining practical knowledge of this HoQ tool is necessary, it might also prove beneficial to understand to what extent the HoQ actually suits and aligns to a CAS. To answer this enquiry, one could perhaps articulate the simplest definition of a CAS as: (a) interactions with elements to achieve a common purpose, and (b) interactions with other subsystems in its adjacent environment to adapt to changes.

In line with such a definition, the HoQ might not satisfy the demand of (b) as far as "physical interfaces" are concerned. However, Figure 5.1 above clearly points towards "competitive interfaces" (i.e., through Competitive Benchmarking) as well as "commercial interfaces" (i.e., the Interrelationship Matrix) to which the new system ought to adapt. Therefore, considering that the Technical Interaction Matrix (i.e., "Roof" of HoQ) addresses the (a) part of avoiding conflict between the interacting elements, it can be safely construed that the HoQ is a suitable "design tool" for large and complex infrastructures when approached as CASoS.

As Tapke et al (2001) further maintain, "The House of Quality helps to identify the critical technical components that require change [...] These critical issues are then driven through the other matrices to identify the critical parts, manufacturing operations, and quality control measures needed to produce a product that fulfils both customer [...] and producer needs within a shorter development cycle time".

The "HoQ for Product Planning" is the first of four "houses" to be constructed. It is the most decisive for it sets the context of the design. In wishing to deploy quality throughout the whole organisation, one would need to involve people responsible for part design, manufacturing process planning, and production planning. QFD is a systematic method that includes all the HoQs, not just the Product Planning House.

The Product Planning HoQ_1 connects the requirements of customers/users to selected engineering characteristics. Then, the Parts Design HoQ_2 connects or relates the engineering characteristics to parts characteristics. The Process Planning HoQ_3 is used to determine process characteristics and/or parameters based on parts characteristics. Ultimately, the Production Planning HoQ_4 will relate process characteristics and production characteristics as per Figure 5.2 (see below). Thus, while HoQ_1 and HoQ_2 determine the design of the new system or the "What to build" (i.e., its architecture), the "How to build" is determined by HoQ_3 and HoQ_4.

Completing the whole QFD process in terms of infrastructure (e.g., LIPs) might still bring about a challenge: the manufacturing processes must be replaced with proper Construction features in HoQ-3 (e.g., concreting, steel erection, piping, walling, finishes) and HoQ-4 (e.g., mortar/concrete mixer, stone crusher, brick maker). Better still, HoQ-4, in particular, should be explored *iteratively* for aspects such as Machinery, Material, Method, Manpower, Milieu or Measurements (6Ms).

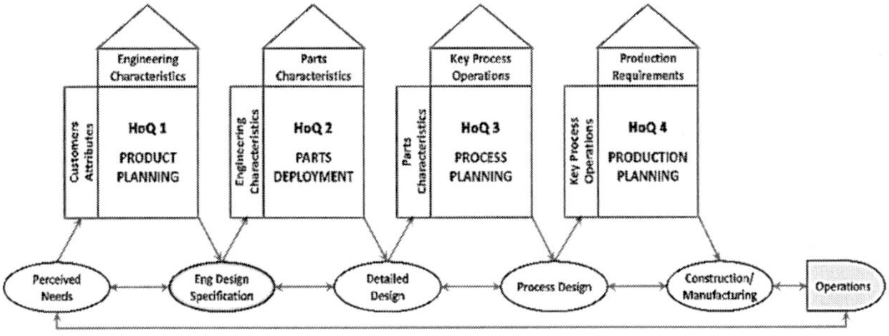

Figure 5.2 – Quality Deployment Function as per Model

5.2 The Functional Analysis System Technique (FAST)

Of the many methods used for System Development, FAST is a concrete and structured approach for Functional Analysis aiding in the determination of the basic functions, critical path of functions, independent functions, and dependent functions, using pertinent *HOW* and *WHY* questions.

FAST is not an end product or result, but rather a "beginning"; it lays open the subject matter under study, forming the basis for a variety of subsequent study approaches and analysis techniques. FAST assists in identifying redundant functions (if any) in a design (e.g., Figure 5.5 exposes redundant functions to be pruned).

Borza therefore argues that, "Too many times, individuals and teams jump into problem-solving activities without fully or properly defining what it is they really need to solve, or what factors or interactions within the problem area will create complications [...] FAST Diagrams provide a graphical representation of how functions are linked or work together in a system (product, or process) to deliver the intended goods or services. By focusing on functions, teams and individuals can focus on what is truly important and not be constrained by physical features of products or processes [and, thus,] opening the creative [engineering design] vista and removing physical and mental constraints in the process" (Borza, 2011).

Most FAST proponents (including Borza) would maintain that its degree of validity is directly dependent on the talents of the participating team members, and the scope of the related disciplines they can bring to bear on the problem. Consequently, the single most important output of the multi-disciplined team engaged in a FAST exercise is *consensus* among the various parties (Borza, 2011).

In a nutshell, the FAST Diagram is the simplified "graphical representation of the underlying functions, requirements, architecture, relationships, and interfaces that define the system" (Jamshidi, 2009). Since FAST determines the "interacting functions" as well as inputs and outputs to/from the environment, it is suitable as a design tool for CAS.

Borza maintains that the FAST Diagram provides the foundation for creating effective function modelling and, thus, for supporting Functional Analysis and Allocation. Figure 5.3 (below) provides a generic depiction of such a diagram (Borza, 2011).

The generic diagram represents both the "leftward flow" and the "rightward flow". The leftward flow relies upon *WHY* questions to determine the functions needed to transform the original "input" (from the environment external to the system) into the ultimate "output" (to the same environment). Conversely, the rightward flow relies upon *HOW* questions to determine the functions required to provide the ultimate "output" (to the environment external to the system) from the original "input" (from the environment).

The convention is that an end or purpose that a "thing" or activity is intended to perform be expressed in a verb-noun form, e.g., "Carry Load". Further, those functions needed to transform "input" or provide the "output" are as follows:

I. *Dependent functions:* Starting with the first function to the right of the basic function, each successive function is "dependent" on the one to its immediate left, or the higher order function, for its existence. That dependency becomes more evident when the *HOW* question and direction are followed.

II. *Independent (or supporting) function(s):* These are functions that do not depend on another function, or method selected to perform that function. Independent functions are located above the critical path function(s), and are considered secondary, with respect to the scope, nature, and level of the problem, and its critical path.

Figure 5.4 (below) depicts the FAST Diagram for the design of a PC Projector system. Considering the intended Higher Order Function (i.e., ultimate output) of 'Share Information' and asking *HOW* questions "rightwards", for example, the flow moves from 'Share Information' to 'Project Image' to 'Transmit Light' to 'Generate Light' to 'Convert Energy' to 'Receive Electricity', and finally to 'Transmit Electricity', which is the Lower Order Function (i.e., original input).

In the same way, Figure 5.5 (below) gives a depiction of a FAST Diagram for the design of an incandescent light bulb, even identifying any *redundant* functions. The designer is, thus, afforded an opportunity to "design out" such functions.

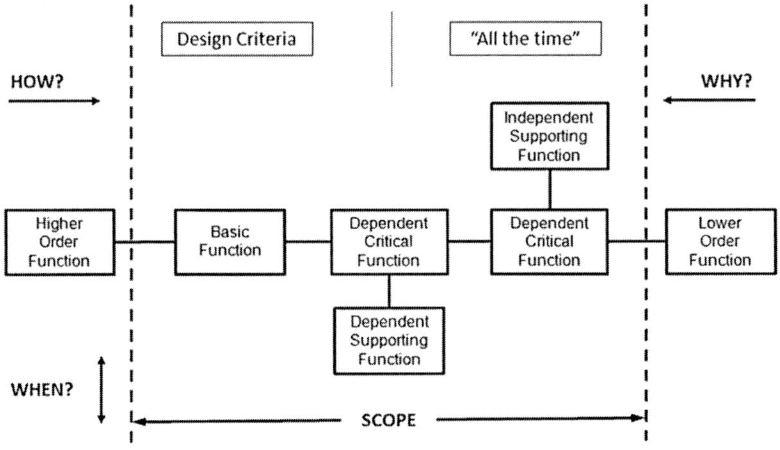

Figure 5.3 – FAST Diagram as per a Generic Template (Source: Borza, 2011)

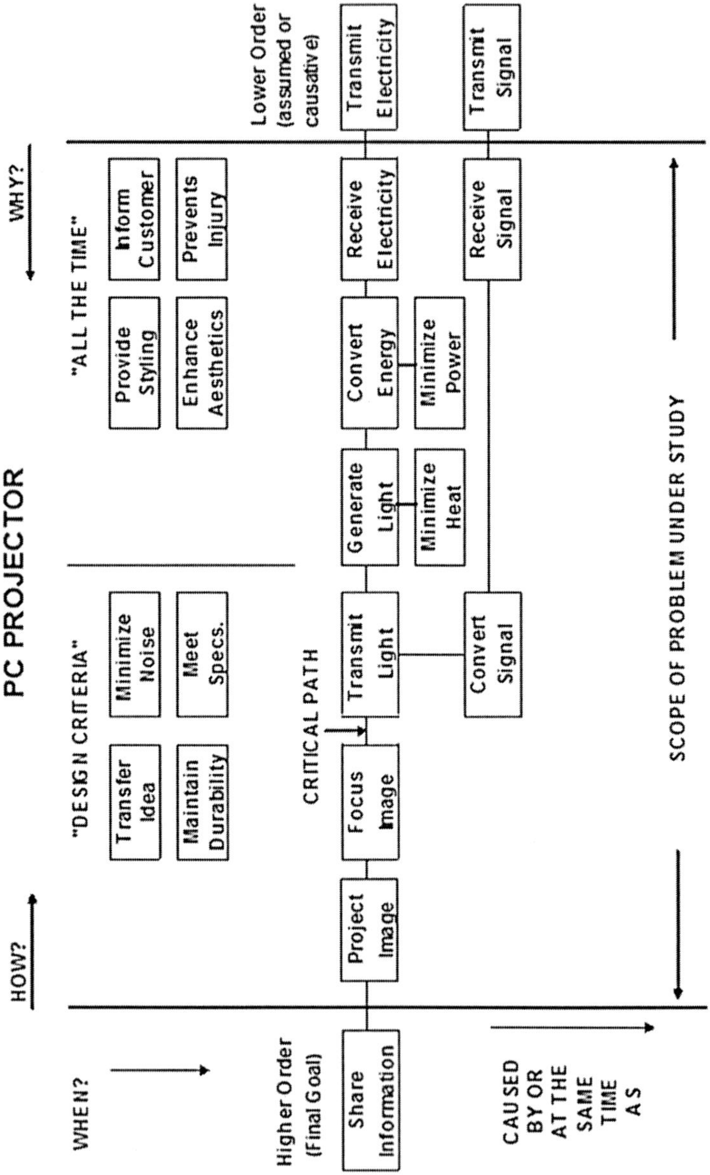

Figure 5.4 – Example of a FAST Diagram for a PC Projector (Source: Borza, 2011)

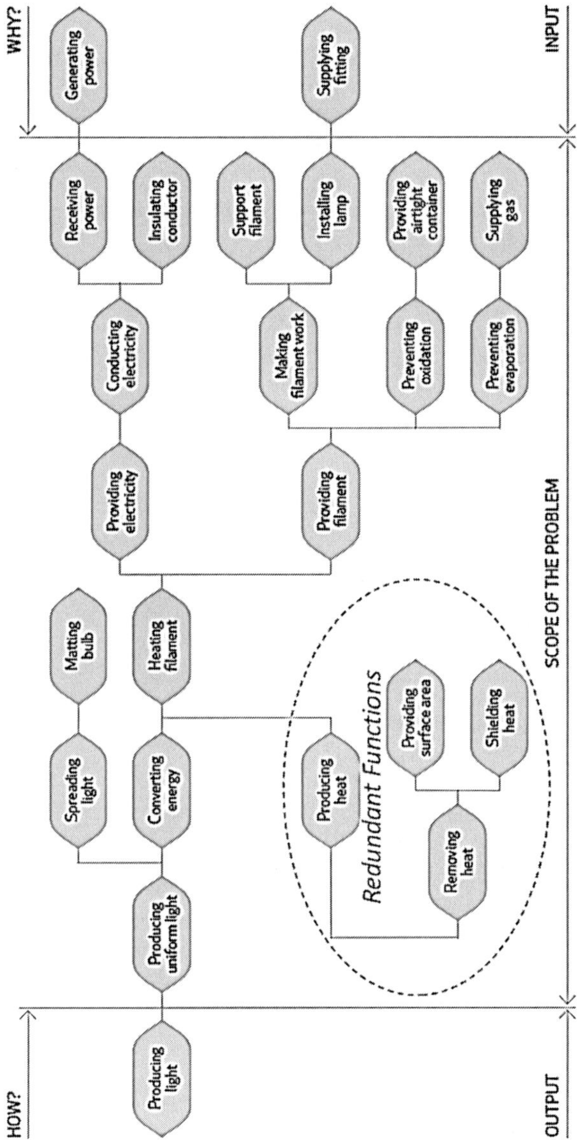

Figure 5.5 – Example of a FAST Diagram for a Light Bulb (Adapted: ProRail, 2013b)

5.3 The Hamburger Model

Large infrastructures such as power plants, water plants, road/rail networks, sea/airports, container terminals, and mega shopping malls are not only complex (i.e., large number of components), but they also need to be "adaptive" in terms of their interactions with the adjacent environment; they are required to be CAS. Further, since their "components" (subsystems, in particular) are also becoming increasingly complex, most mega-structures are nowadays proving to be CASoS.

Therefore, a system-based method/tool is needed to design around these "components". The Functional Unit (FU)–Technical Solution (TS) or the FU–TS Model separates each building-object within a subsystem into a functional appearance and its technical implementation. The FU collects all the information (e.g., requirements, functions, logics, basic assumptions) required to make a choice for a TS (i.e., with characters to be verified), according to the Dutch MIE (ProRail, 2013).

On that note, Van Ruijven (2013) maintains that:

> "The so-called 'Hamburger model' principle (FU–TS Model) is used as a base to build the information models [...] This concept makes a clear distinction between a certain abstract object (represented by e.g., a symbol on a drawing) and the materialized version of it (as delivered by a supplier and installed in place). The FU can be seen as representing all the requirements (functional, technical, performance, constraints and other aspects including, e.g., [cost/price] [,] safety, esthetical and maintenance). An FU captures a 'design problem' and 'exists' in principle as long as the life cycle of the total system lasts".

Van Ruijven further recommends that, "The Technical Solution can be a principle or technology or a tangible piece of equipment, described with a set of characteristics which fulfil all the requirements of the FU. The selected Technical Solution can be replaced more than once during the life cycle of the total system. In the presented approach is [the idea that] an FU can be either a Functional Object or a Functional Physical Object in the context of ISO 15926" (Van Ruijven, 2013).

The author was once asked: "*Why the hamburger in the model seems empty? Why is there no 'filling' of meat, cheese or lettuce in it?*". He jokingly, but correctly answered that the '*meaty filling*' shall come at a later stage from the Detailed Design process—the hamburger shall be complete at that stage! Indeed, the hamburger model usually appears at the early Conceptual phase.

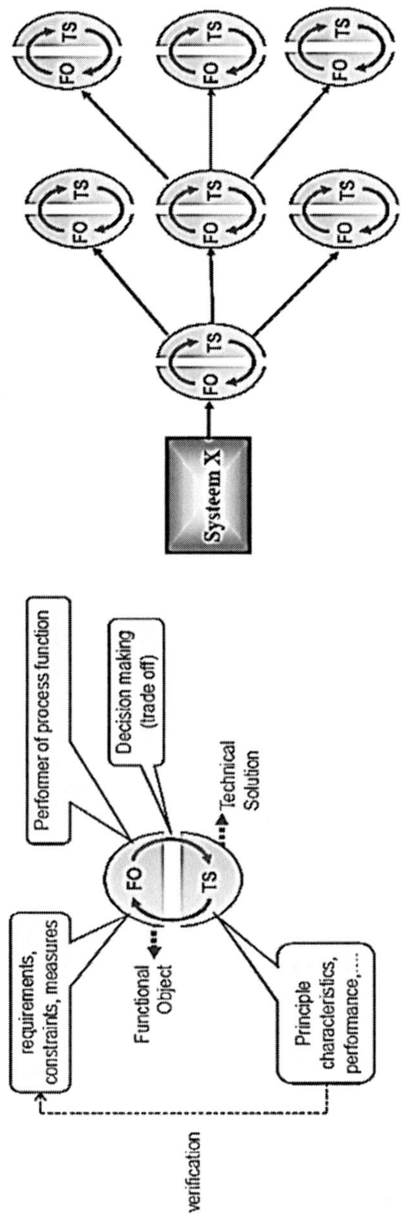

Figure 5.6 – The Functional Object – Technical Solution Model (Van Ruijven, 2013)

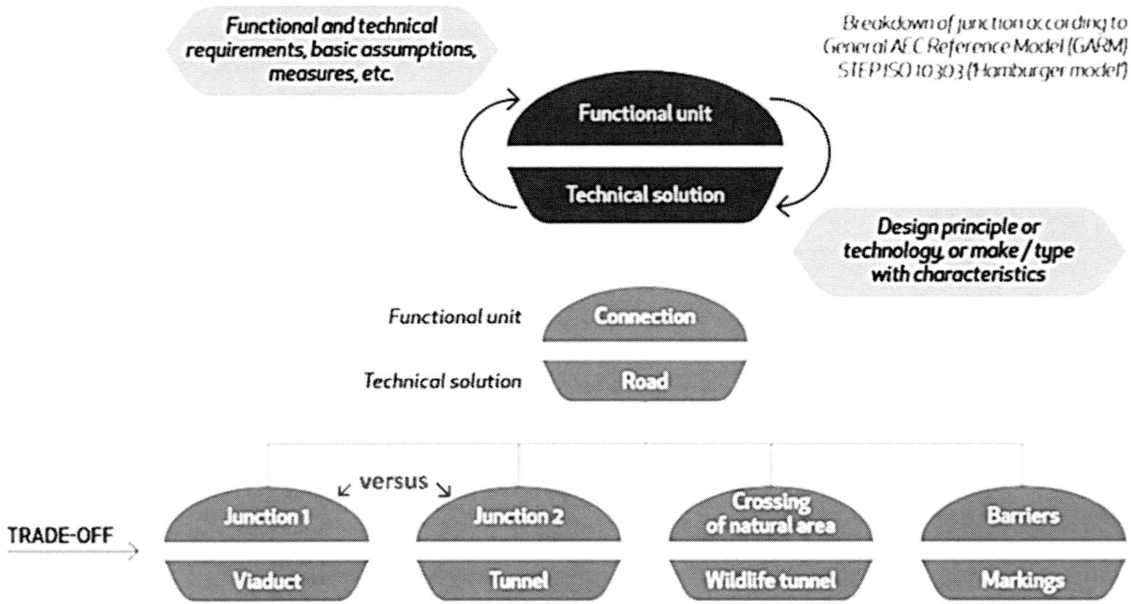

Figure 5.7 – Hamburger Model for the Design of Road Connection (ProRail, 2013b)

Designing a new infrastructure from a "components" perspective is in line with the very definition of systems. The whole approach also entails "*breaking the envisaged system into its components*". Hence, "Based on the Hamburger model one can build a hybrid System Breakdown Structure representing the Functional Objects, design principles and or Technical Solutions on the various levels of the breakdown structure [see Figures 5.6 and 5.7]. This approach doesn't require an additional functional breakdown with a mapping to the physical breakdown structure anymore as traditionally is done" (Australian Government, 2012).

While the structure of the infrastructure system will result from the System Breakdown Structure (SBS), due care should also be taken in ensuring that their collective behaviours achieve the stated purposes. "The emerging behavior of an SoS is the direct result of *synergy* among its constituent systems, interaction of each constituent system with its immediate surroundings, and the interaction between the SoS and its environment [...] Technical planning for an SoS will need to augment as well as take into account the plans of those constituent systems" (Jamshidi, 2009).

"It must also be initiated top-down but iterated with the constituent systems until a consensus (but not necessarily optimum) solution is agreed upon and resourced [...] The SoS architectural analysis and development is an iterative process of elaborating, partitioning, decomposing, and allocating the SoS functional and performance requirements across the constituent systems [and their subsystems]" (Jamshidi, 2009)—otherwise the desired overall behaviours will fail to "emerge", almost like a *flopped* cake dough!

Figure 5.7 below gives a depiction of a Hamburger Model for the design of a road connection. This particular infrastructure will consist of components such as 'Barriers', 'Crossing of Natural Areas', 'Junction 1' or 'Junction 2'. It is expected that for each "hamburger", a set of functional/technical requirements shall match a specific "component" (i.e., technical solution) that meets the design criteria. Furthermore, a trade-off exercise is expected to take place between FU and TS aspects or between, for instance, subsystems that serve substitutable purposes.

The simplicity of this "Hamburger/FU–TS Model" could be a bit deceiving. It is a potent tool for the design of a large and complex infrastructure when viewed from a "components" perspective. Nevertheless, the model has a downside in that, it shall be said, the FU–TS Model often fails to identify *beforehand* which emergent behaviours from the interactions between those interacting components might prove "undesirable".

This Hamburger/FU–TS Model has been effectively applied by NASA engineers to the design of the Space Shuttle, which is "*a combination of specialised parts*" such as the 'Orbiter', the two 'Rockets Systems', and the 'Orange Tank' that stores liquid fuel to constitute the most complex (i.e., a re-usable spaceship) and one of the most expensive machines ever built and capable of flying at more than 25 times the velocity of a speeding bullet.

At the next level of its Functional Breakdown Structure (FBS), leading to the SBS features, are three 'Rocket Engines' that are attached at the rear of the 'Orbiter'. In themselves, they burn fuel at 3,300 ° C to produce 37 million horsepower. There is also the highly instrumental "six-joint" 'Robotic Arm' that corresponds roughly to the joints of a human arm, which is mounted to handle payload weighing up to 29,000 kg (Galvez et al, 2010).

This Hamburger or FU–TS Model is surely a powerful design tool that engineers might have been using all along without even realising it. For instance, the author himself was once appointed to lead a medium project involving a "Training and Testing" facility for a mining company. The scope entailed refurbishing a disused workshop and converting it into a dedicated training facility for junior technicians.

What could have been a mundane (if not borderline boring) assignment turned into a challenging enterprise owing to one specific requirement: "*For safety reasons, the roof of the facility shall remain open during 'blue testing', but tightly closed during 'red testing' or similar*". The main difference between blue and red testing was that the former (i.e., blue) entailed "replica diamonds" and caused fumes, while the latter involved "real diamonds" but emitted almost no fumes during the process.

Having developed an ordinary Work Breakdown Structure (WBS) of the facility, the team's attention turned and focused heavily on the 'roof' package that then became a *make-or-break* element of the entire project. The question arose in terms of who (i.e., suppliers/contractors) out there will be able to design, manufacture, and install *"a roof that can open and close (tightly) at will"*—and do so both safely and effectively?

The project team visited and interviewed a number of contractors and suppliers, including sliding gate and roller-shutter door manufacturers to provide the technical solution to that *sine qua non* requirement. Unless and until the right "technology" is found that meets the technical and functional requirements, the otherwise high-priority project would have to be aborted.

At that time, the proposal of "*a foldable steel trussed-roof sliding on rails*" had met the structural stability requirements but was not deemed "safe enough" (e.g., tightness, steal-proof) by the client, so the project stopped. For until the Hamburger (FU–TS) Model works and proves satisfactory, there will be no system that works.

5.4 The N-squared Diagram or Design Structure Matrix (DSM)

The allocation of System Requirements to pertinent subsystems is critical in determining the architecture of the system under consideration. It is this specific arrangement of subsystems (i.e., system architecture) that will usually govern the overall behaviour of the envisaged system. Indeed, the behaviour of a system is profoundly driven by the *interconnectivity* of the components (i.e., subsystems) that constitute the architecture of that system.

One of the roles of the systems engineer is to examine the different ways components can be aggregated into subsystems and use a design approach that maximises cohesion and minimises any unnecessary coupling. This contributes to optimising those *useful* interactions between the interfaces of the subsystems. A suitable analysis instrument is thus needed that can help identify the "natural" interconnectivity that exists (or ought to have existed) in all systems and subsequently exploit this in determining the most appropriate architecture design and, alternatively, to analyse and elucidate the general or overall behaviour of an existing or "planned" system.

A useful tool to perform this activity is the N-squared Diagram (Kasser, 2015). The tool was invented by Robert Lano in 1977. In its general form, it is a table that shows the relationships between parts, entities, functions, people, equipment, organisations, etc. Therefore, the N-squared Diagram is an analysis tool that uses a $n_x n$ (n by n) matrix to record the interconnections between elements of a system.

In terms of System Design, the N-squared Diagram is used to assess the degree of binding and coupling in a system and thereby to determine the candidate architectures based on the natural structure of the system. In addition, the tool may be used to record and, therefore, aid the management of interfaces in the system being considered. Furthermore, in System Analysis, the same tool could be used to identify (and, thus, document) the interconnectivity in a system and help understand observed behaviours and provide guidance for improvements.

Whereas a given System Breakdown Structure (SBS) identifies where these interfaces occur, the N-squared Diagram depicts relationships between system entities and relates the entities to the SBS. In this manner, System Integration could take place at every level of the system architecture. Hence, the entities are listed on the diagonal of the matrix, with outputs shown in the rows and inputs in the columns (Forsberg et al, 2005).

The N-squared Diagram or Chart is a graphical depiction of the functions within a system, together with the one-way interactions between each function ordered in a (n x n) matrix, with function or entities on the centre descending in diagonal cells. Since functions occupy each diagonal cell, the total number of cells is equal to the square (i.e., N^2) of the number of functions, hence the name. Interface functions and constraints are shown in the cells that correlate to both interfacing entities and the graphic is read clockwise, from left to right (Forsberg et al, 2005).

When applied in Systems Design, the size of the matrix shall be equal to the number of system elements (i.e., subsystems) or functions. Interconnections between functions or system elements are captured in the off-diagonal cells. By convention, however, outputs are in rows and inputs are in columns. Inputs from outside the system, and outputs from the system are shown in line with their respective source or destination function or element.

The N-squared or N2 Diagram is used to develop system interfaces. The system components or functions are placed on the diagonal; the remainder of the squares in the matrix represent the interface inputs and outputs. Where a blank appears, there is no interface between the respective physical components or functions. The N-squared Diagram could be taken down into successively lower levels to the component functional levels (e.g., subsystem). In addition to defining the interfaces, the chart pinpoints areas where conflicts could arise in interfaces and highlights input and output dependency assumptions and requirements—this is a key issue in preventing *non-synergistic* connections (Systems Engineering Handbook, 2007).

Considering that “The greatest leverage in system architecture and design is at the interfaces [i.e., where two or more systems/components interact]” (Maier and Rechtin, 2009), the N-squared Diagram is a valuable tool for not only identifying functional or physical interfaces in a system, but also for identifying areas and components in which conflicts and *dysfunctions* (i.e., bad behaviours) may arise with interfaces so that “system integration” may proceed smoothly and efficiently.

According to Figure 5.8, [*Component 1* + *Component 2*] constitutes a natural subsystem; yet [*Component 2*] alone is a "Critical Element" as it provides input to every other component and *conversely* receives output from any other component. A failure of this particular component will cause the complete failure of the system.

The N-squared (N^2 or N2) Diagram or Chart flows from the concept map, which is a view of the functions involved in a system and their interactions. However, the concept map is not very useful for the purpose of depicting the aggregation of subsystems since it is difficult to see any useful pattern in the interconnections between the entities. It is only when the functions are drawn in the form of the N^2 Diagram that patterns would be seen in the interfaces.

Component 1	X		X	
X	Component 2	X	X	X
	X	Component 3		X
X	X		Component 4	
	X	X		Component 5

Figure 5.8 – Example of an N-squared Diagram or Chart

An example of an N-squared Chart could be developed for the following "Coffee Machine" system which is purposely kept straightforward for ease of illustration:

> *"Coffee Machine [as a system] that serves coffee at the right temperature; with the correct amount of sweetener; with the correct amount of milk; and, with the desired strength of coffee."*

In an actual instance of EDD, this endeavour should have started with the development of Causal-Loop (CL) Diagram(s) that would have determined cause-and-effect relationships in the given scenario. Such CL Diagram(s) would then evolve into a Function Flow Diagram (or into a Context Diagram), the purpose of which is to guide the devising of a set of systems intended to solve the problem (in this specific case, create a "*Coffee machine that serves coffee*").

Due to space limitation, the author has proceeded straight to draw a Function Flow Diagram. Figure 5.9 below captures the Function Flow Diagram for the planned "Coffee Machine" (i.e., as a system), which then informs the ensuing N^2 Diagram.

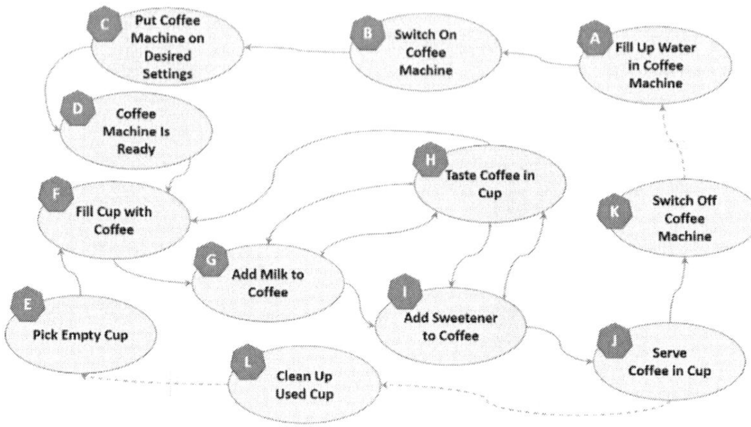

Figure 5.9 – Function Flow Diagram for a "Coffee Machine"

It would appear from the Function Flow Diagram for the "Coffee Machine" that a CL Diagram should have informed the cause-and-effect relationships between the G-H-I (i.e., "*Add Milk"; "Add Sweetener; "Taste Coffee*") functions of the system.

Furthermore, Figure 5.10 identifies the *natural* interconnectivity between the required functions and, thus, candidate subsystems of the "Coffee Machine". The architecture of the "Coffee Machine" system could be re-configured from its initial arrangement as A-B-C-D-E-F-G-H-I-J-K-L to an aggregated [A-B-C-D]-[(E-F-G-H-I)(I-J-L)]-K.

The latter architecture (as reconfigured) is far simpler (i.e., four subsystems) despite containing some highly complex subsystem, the E-F-G-H-I. Indeed, there are practical benefits in devising a module that can "*pick cup; fill cup; add milk; add sweetener; and taste*". This is where the automation resides.

The same architecture would also suggest that sub-systems E-F-G-H-I and I-J-L may be combined or integrated to carry out the composite E-F-G-H-I-J-L function. Further, while element I may appear to connect subsystems E-F-G-H-I and I-J-L, an analysis of E-F-G-H-I's internal structure shows that element H plays a very crucial role in *interfacing* these subsystems—since element H proves such a critical interface for G-H and H-I components. This remarkable *insight* implies that should module/function H fail, not only will E-F-G-H-I fail, but the whole system as well—and the whole automation (of coffee service) will turn futile, if not detrimental (i.e., poisoned chalice).

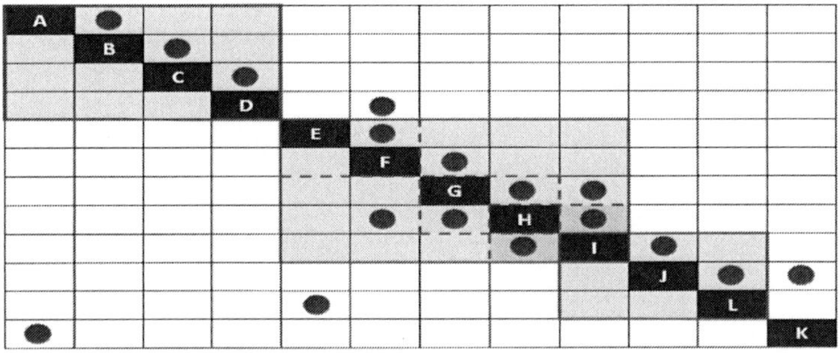

Figure 5.10 – N-squared Diagram for a "Coffee Machine" System

In arguing the "*Seven Principles*" for systems engineered solution systems, Kasser (2015b) insists on maintaining reliable, effective "interfaces" when designing subsystems:

> "Each subsystem [e.g., the E-F-G-H-I unit] may be developed independently and in parallel with the other subsystems provided that fit, form, function and interfaces are maintained throughout [...] it is necessary to have budgets for the whole system, as well as budgets for each of the subsystems [and their interfaces]." (Kasser, 2015b)

While this basic example of a "Coffee Machine" might look trivial, this approach would prove powerful in allocating functions to natural candidate "subsystems" that govern the architecture and, thus, the general behaviour of large and complex systems such as airports, water/power plants, chemical plants, hospitals, schools, and shopping malls.

The author has proceeded to retrospectively apply this method to analyse the architecture of a campus at a particular university, following its massive expansion of late. The campus started off with a street-facing administrative building surrounded by a number of faculty blocks and has since expanded to accommodate not only more faculties, but also more students and, consequently, more daily visitors.

In response to a growing need for parking (as exacerbated by the near collapse of the public transport systems) the university has multiplied its parking capacity, converting every open space into a parking lot. While they managed to reduce parking deficits in terms of absolute numbers, the scattering of parking spaces to many locations (with no "*Parking is Full*" signposts) has created unnecessary traffic with visibly annoyed drivers roaming to and from various parking hubs, seeking an "open" parking bay.

This acute predicament (which has a lot to do with "campus architecture") has proved more hurtful than the actual shortage.

Students and lecturers themselves are not spared either. For instance, a trip from the science precinct (where some lectures are held) to the chemistry labs can take up to a 10-minute walk, at times under rain or snow in inclement weather. Further, faculty-based libraries and computer pools are usually too full for the intended users, forcing them to take a 15-minute walk to the nearest central facility. Worse still, medical school students have a 15-minute bus trip from the medical school (actually located in another suburb) to attend shared classes on the main campus.

Moreover, many "centralised" support-services that are in daily demand (such as ICT support, security, restaurants, and the post-office) are still located in the old administrative building, which is some miles away from certain faculty blocks. There is no reason for having these services close to the vice-chancellor's office, other than a short-sightedness as far as this campus' architecture is concerned. These support services, in particular, ought to have been relocated to a more central location considering the expanded footprint, but they surely were not.

A university is a system of "teaching and learning". Nevertheless, such operations shall not be impaired by a flawed architecture that makes its campuses "*behave badly*" due to poor aggregation of facilities and capacity allocation. This usually engenders an overall behaviour that is not conducive to teaching and learning. It is not only about a university; many mega-airports do suffer a similar malady—they might actually need *surgery* to alleviate/eradicate their current disfunction.

5.5 The Context Diagram

System interactions occur at the "interfaces" between systems, between subsystems, and between subsystems any other system elements. Those interactions involve input-output relationships, function tracking and other aspects that play a crucial role in allowing the infrastructure to operate as a Complex Adaptive System (CAS).

When presented with an engineering design problem, the system engineer's *first* task is to truly understand the problem in its totality; this means understanding the "context" in which the problem is set. Once the system components are defined (using FU—TS Model, the N-squared Chart, and others), a block diagram showing the major components, interconnections, and external interfaces of the system should be developed to define the components and their interactions (Systems Engineering Handbook, 2007). Such an approach is entirely consistent with both Systems Thinking and SE.

NASA engineers rely upon the Context Diagram as a tool for interface design. A Context Diagram is a useful tool for grasping the "soul" of the system and the "external domains" that are relevant to and have interfaces with such a system.

It depicts the general structure of a context diagram whereby the system is shown surrounded by the external systems which have interfaces to the system. These systems are not part of the system being studied; they interact with the system via its external interfaces. Those external systems can impact the system, and this does have an impact on the external systems. As such, they play a key role in elucidating the requirements for the system at hand (Systems Engineering Handbook, 2007).

Therefore, "We define interfaces in a system design to minimize unintended [and probably undesirable] interactions and simplify development and operations and then we document and control the design [...] engineers who are developing an avionics system [or any other large and complex systems] must practice creative design and interface definition to achieve their goals" (Ryschkewitsch et al, 2009).

The foregoing discourse will suggest that for the purpose of this section, the focus should rather be on "external" types of system interfaces—with adjacent systems. Based on their extensive and insightful experience, NASA engineers argue the need for managing interface requirements as follows:

> "It is important to define all interface requirements for the system, including those to enabling [as well as sustaining and competing] systems. The external interfaces form the boundaries between the product and the rest of the world. Types of interfaces include operational command and control, computer to computer, mechanical, electrical, thermal, and data. One useful tool in defining interfaces is the context diagram, which depicts the product and all of its external interfaces." (Systems Engineering Handbook, 2007)

A good example of a Context-Diagram is provided below in Figure 5.11.

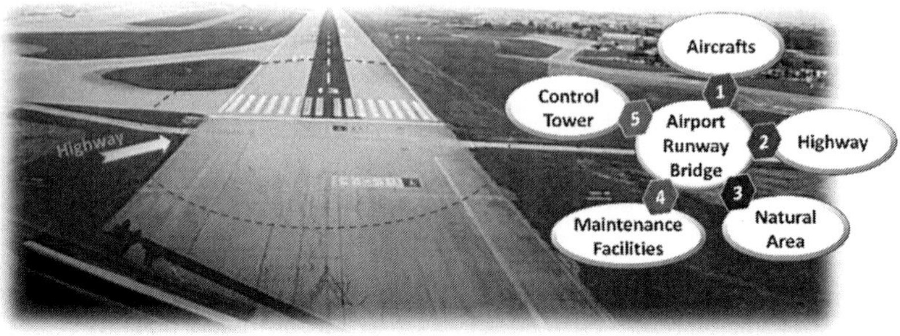

Figure 5.11 – Example of a Context Diagram for an Airport Runway

#	Nature and Requirements of the Interface
①	The Airport Runway Bridge is designed for larger aircrafts such as Boeing 777-300 and Airbus A380 (that occasionally visit for repainting); they need a longer runway to land at this particular airport. The structure and dimensions of the Bridge shall allow for landing of larger planes; for instance, a fully loaded Airbus A380 may weigh up to ± 590 tons.
②	A Highway is to continue underneath the Runway through the Bridge. The Highway is to have suitable clearance for planned traffic (e.g., trucks); lights and barriers shall be installed that conform to aviation codes.
③	The Airport Runway Bridge is located in a natural area and shall fit in with its layout; native plant seeds are to be sown along paths, on verges. The direction of the runway shall conform to airport operations in terms of wind and/or magnetic directions as well as safety codes.
④	The Airport Runway Bridge is to connect and direct traffic of occasional aircrafts visiting for maintenance/repairs to appropriately designated and located maintenance and other painting facilities.
⑤	The Airport Control Tower shall direct larger aircrafts (e.g., Airbus A380, Boeing 777-300) to land "*beyond-highway*", using the "extended" landing Runway while any smaller aircrafts will land "after-Highway".

Table 5.1 – Legend on Interfaces Captured in Figure 5.11

The Context Diagram is appropriate as a tool for effectively designing those interfaces, both those impacted and those impacting the system under consideration. Once interfaces are fully recognised, then requirements and assumptions can be described and tested at critical moments, which assists in preventing the sub-optimisation of the several disciplines appointed to "engineer" the various systems, subsystems and assemblies of the "system" being considered. Kasser concurs:

> "The context diagram is shown in Figure […] where the system being acquired is the spiked object in the centre. The system has an interface with several other adjacent systems but not necessarily all the systems in the [given] framework." (Kasser, 2013)

Jamshidi elevates this discourse to the realm of SoS, which relates to the reader's inquiry on large and complex infrastructures. He submits that, "An engineered SoS should be designed and developed as an open system [i.e., with its external interfaces] in order to continually exchange material (e.g., constituent systems, hardware and software modules, resources), energy (e.g., power and signals), and information (e.g., meaningful data) with the environment encompassing it". (Jamshidi, 2009)

Van Ruijven (2013) also elaborates on the significance of interfaces as follows: "Physical objects interact with their environment, human or other system elements by means of an interface which consists of one or more ports. All interactions that occur during the operational life cycle stage of the system can be defined using this port principle [...] There are four basic types of ports: material ports, energy ports, information ports and construction (3D) ports [which inform the "How to build"]. Interactions by means of a port can take place between functional objects and between functional physical objects and the environment and/or stakeholders. An interaction with corresponding [interface] ports becomes a physical connection when a functional physical object is realised by a materialised physical object".

The five external interfaces indicated in Figure 5.11 regarding the Airport Runway Bridge are further discussed in Table 5.1 for any detailed reference purposes.

In defining the interfaces and relevant exchanges, the engineering team gains insights into the "design and manufacturing/construction" of the required ports. "Within these projects, all project partners (multidisciplinary) use and share the same way of defining a risk, interface, [...] the SBS and the WBS which improves significantly the interoperability within these projects. The information models also enable product manufacturers [i.e., contractors] and/or suppliers to create 'product knowledge models' of their products in such a way that product information can be easily integrated in a systems engineering environment based on the same information models" (Van Ruijven, 2013). Many megaprojects have efficiently applied this notion.

For instance, the 2,880 m long, 28 m wide Rion-Antirion Bridge (Greece) is a creation of extraordinary engineering—i.e., built in an area with high seismicity and having the longest continuous, cable-stayed, fully suspended deck in the world, measuring 2,252 m in length (Pararas-Carayannis, 2020). New engineering techniques (in design and construction) had to be devised.

The bridge has been built to withstand the collision of a 180,000-ton oil tanker travelling at 16 knots (a horizontal impact load of up to 28,000 tons), winds of 250 km/h, deck movements from seismic activity of up to 2 m in all directions between two adjacent piers and tectonic fault spreading of approximately 1.6 cm per year (2-5 m in 125 years). Lastly, with the incorporation of a stringent design for seismic loading, the bridge is able to withstand sudden ground accelerations of up to 0.5-g, maximum spectral accelerations equal to 1.20-g (from 0.2 to 1.0 seconds) and earthquakes of magnitude 7 on the Richter scale.

Specific engineering designs were used in bridge construction to mitigate the dynamic effects of thermal and/or rift expansion, windstorm damage, tilt, and seismic activity at *critical* interfaces (Pararas-Carayannis, 2020).

This approach applied to three physical objects (Pararas-Carayannis, 2020):

I. "Foundation construction techniques to mitigate the absence of bedrock, effects of aseismic or seismic tectonic movements, rift expansion and tilt." (Ibidem)

II. "Pylon construction that reduces the effects of wind, seismic activity, rift expansion, asymmetrical service loads, aseismic tectonic movements, etc." (Ibidem)

III. "The main bridge and deck with fluid viscous 'dampers' and fuse restraints to overcome unique challenges, e.g., mitigate the effects of wind, aseismic tectonic movements, seismic activity and rift, and thermal expansion." (Ibidem)

These innovative, precautionary designs made it possible for the Rion-Antirion Bridge to stand where no other could have stood, and perhaps do so for a long time having survived the 5.3 magnitude quake that jolted central Greece on 30 March 2019 (with an epicentre in the Gulf of Corinth, right where the bridge is located).

Just as innovative designs of critical interfaces have allowed the Rion-Antirion Bridge to overcome the threat of earthquakes in Greece, some 9,500 km away in Japan, the same engineering approach made it possible for the *Shinkansen* ("Bullet Train" in Japanese) to reach an *unbelievable* maximum operating speed (MOS) of 320 km/h (200 mph) on a 387.5 km section of the Tōhoku Shinkansen, with test runs having reached up to a 603 km/h (375 mph) world record for SCMaglev trains—the awaited magnetic-levitation version in April 2015 (Hood, 2007).

The secret of what makes the Shinkansen "fleet-footed" is hidden at its "feet". "Shinkansen uses a range of advanced technology compared with conventional rail, achieving not only high speed but also a high standard of safety and comfort" (Hood, 2007). Such technologies are located at its critical interfaces. Shinkansen designers figured out that by flattening the "conicity" of the wheels and by introducing coil-suspension systems and computer-controlled airbags, they could eliminate the "hunting oscillations" (or wobbling) of the bullet train.

The engineers also introduced "continuous-welded" rail and swingnose crossing-points, eliminating gaps at turnouts and crossings. Long rails are used, and are joined by expansion-joints to minimise gauge fluctuation due to thermal elongation and shrinkage. Further, a combination of ballasted and slab track is used, with slab tracks (consisting of rails, fasteners and track slabs with a cement asphalt mortar) being exclusively used on concrete bed sections such as viaducts and tunnels (Hood, 2007). Prefabricated upstands prevent the track slab from moving or shifting along either the latitudinal or longitudinal directions, thus making the ride smoother.

With weight-reduction and further improvements on pantographs, power is distributed along the axles of the train to reduce the heavy axle loads under single power cars. *Shinkansen* trains are electric multiple units, offering fast acceleration, deceleration and reduced damage to the track due to the use of lighter vehicles compared to competing types of locomotives or power cars.

Then *voilà* — welcome to the 320 km/h smooth ride!

5.6 The Lean Manufacturing

Eduardo argues that a self-organised project team should "*return to a zero-information state*" (Eduardo et al, 2000). Precisely, this idea of "*return to a zero state*" shall concern itself with former knowledge, not with the needed input-resources.

For Statistics from the US National Institute of Standards and Technology (NIST, 2006) suggests that "about 30% of project costs can be attributed to waste: waste in information transmission, waste in rework and waste in non-productive activities within the project [and within systems]" (Krauss, 2014). Unfortunately, particularly in the LIPs industry, many design engineers would start off their design endeavour as if any new or novel systems *ipso facto* shall entail using new, off-the-shelf materials, parts, subassemblies, subsystems, and other consumables.

This stance is divergent to the sad reality of our planet already facing an Earth Overshoot Day; meaning, the day our society has consumed more resources than our planet is able to replenish for that particular year, e.g., on 29 July in 2019. This is no surprise at all since during the 20th century, global raw material use rose by twice the rate our population grew. Worse still, raw material consumption grew by 1.4% for every 1% of GDP growth. Hence, depletion of key resource is fast approaching and, given their large scale, infrastructure projects are one of the obvious culprits.

For these reasons, the author previously wrote, "Improvements to the operational environment should involve Lean Operations: Lean Thinking is a holistic paradigm that focuses on delivering maximum value to the customer […] minimizing wasteful practices [...]", "with perfect quality and no waste" (Mabelo, 2020b)—system designers shall pursue "*elimination of waste*" during any EDD activity.

Jamshidi takes a similar perspective on LIPs—when construed as CASoS:

> "Additionally, plants' waste is food for other plants or organisms […] In engineering systems [e.g., LIPs], man has not yet been able to remove the boundaries between resources and waste. Systems engineering and development has still a long way to go in minimizing waste and conserving energy and [input] material. To conserve resources, SoS [design] engineers should consider all the interrelationships and interdependencies that exist within the SoS environment and among its constituent systems to identify the sources of waste in an SoS and formulate strategies to conserve [input] material and energy through an open systems design." (Jamshidi, 2009)

The question is: "*How to achieve such conservation or waste elimination?*" To answer this interesting question, the author would rather refer to Slack:

> "While any operation's current performance may be far removed from such ideals, a fundamental lean belief is that it is possible to get closer to them over time. Without such beliefs to drive progress, lean proponents claim improvement is more likely to be transitory than continuous. This is why the concept of continuous improvement is such an important part of the lean philosophy. If its aims are set in terms of ideals which individual organizations may never fully achieve, then the emphasis must be on the way in which an organization [i.e., as a *designed* operation] moves closer to the ideal state. The Japanese word for continuous improvement is Kaizen, and it is a key part of the lean philosophy […] Arguably the most significant part of the lean philosophy is its focus on the elimination of all forms of waste." (Slack et al, 2010)

The LIPs industry may well borrow this "improvement approach" from the field of manufacturing. Lean Manufacturing can be defined as a systematic method for the elimination of several types of waste. "Lean" methods entail "*the removal of operations [e.g., functions, behaviours, and components] that do not add value within a manufacturing [or infrastructure] system*" (Araghi, 2016). Therefore, LIPs should be designed to secure the provision of certain services or of certain products "in ways that remove or diminish waste" (Araghi, 2016) as follows:

Muda is any activity/task that does not add (or rather destroys) value in terms of:

- Transport:
  The unnecessary movement of a product between steps/activities of operation
- Inventory:
  The volume of "stored" raw materials, work in progress, and stocks of finished goods
- Motion:
  The unneeded, misplaced physical movement of persons or machines in conducting tasks
- Waiting:
  The idle time or waiting for prior activity to be complete or product to arrive
- Over-processing:
  The conducting of operations beyond those required by customers or standard

- Over-production:
  The redundant production of excess inventory, product, information or energy
- Defects:
  The excessive rejection of products or parts that do not conform to standards
- Skills:
  The failure to effectively utilise the skills and/or knowledge of staff and/or employees

Mura is the type of waste that manifests due to inconsistencies and/or unevenness in operations (e.g., when demand is not smoothed out or unfair demand is placed on the organisation's processes and people). Therefore, addressing Mura generally entails incorporating a degree of flexibility in the system's design (Early, 2016).

Muri (i.e., overburden) is the type of waste which arises from failures in system operations, e.g., lack of training or standard operating procedures (Oehmen, 2012), as well as applying incorrect tools and improper resource allocation (Early, 2016).

As far as the author is concerned, any "*improvements in operations*" shall start at design. "Lean Thinking [as applied to Engineering Design and Development] is the dynamic, knowledge-driven, and customer focused process [i.e., starting right at design] through which all people in a defined enterprise continuously eliminate waste with the goal of creating value" (Mabelo, 2020b)—it entails the removal of tasks, activities, functions or components that do not add value to the system or operation.

In the context of large and complex infrastructures that should be designed as a CASoS, "Lean" principles such as the "Mura", "Muda" and "Muri" will be applied in order to eliminate waste from processes—even making sure that "*unnecessary work is not done*":

(i) Within the system, appropriate "Lean" principles shall be applied between "components" to eliminate or minimise waste not only by designing out any potential waste in their respective interactions (e.g., unnecessary friction, energy loss, information asymmetry), but also by possibly reusing appropriately selected parts, subassemblies, subsystems and raw materials from previously operated or disused systems.

(ii) Similarly, "Lean" principles should be applied between the system and its surrounding environment in order to concurrently reduce the number and/or volume of consumables (e.g., lubricants), needed for the system to operate effectively and the amount/toxicity (e.g., filters and exhausts) of discharges back to the environment—to minimise its waste footprint wherever practical.

Engineering design should therefore explore reusing "recycled" items such as steel reinforcing bars, crushed rubble, and equipment (e.g., HVAC). Any innovative design that alleviates friction between interacting parts (and, thus, decreases the needs for lubricants) would be appreciated. In the same vein, those construction methods and/or processes that reduce reworks (for example, the amount of rubble) must be encouraged and promoted. Surely, such objectives are easily achievable.

For instance, the city of Delhi alone generates 4,000 to 5,000 tons of "Construction and Demolition" (C&D) waste per day, contributing to more than 20% of solid waste pollution in the metropolis. Infrastructure building and related activities are the major source of C&D waste in India. Nevertheless, the North Delhi Municipal Corporation Initiative (MCI) effectively recycles 95% of its 2,000 tons per day capacity to re-usable construction materials (i.e., bricks, tiles, sand, crushed stone).

Kasser encourages this attitude (i.e., waste elimination and/or recycling) right from early stages (i.e., design), noting that waste generation is intrinsic to any function:

> "A function by definition transforms an input to an output while consuming resources and generating waste [thus, waste generation is a given]." (Kasser, 2015)

> "One can see that there are two system outputs, the desired output labelled 'output' and another [bad] output labelled 'waste'. If applied to the incandescent light bulb, the desired output is light; the undesired or waste output is heat." (Kasser, 2015)

> "Process improvement and cost reductions efforts tend to focus on the effectiveness of producing the desired output, namely the mission functions. The template shows that there is also a need to focus on reducing the amount of waste or exploring ways in which 'waste' can be used or treated as a desired output and sold to someone who wants the waste. For example, waste heat might be sold or used; waste food in a restaurant might be used for fertiliser or animal food and so on." (Kasser, 2015)

> "Lean systems engineering takes place when the unique systems engineering process designed to realize the system does not contain any non-productive activities. Since non-productive [i.e., non-value adding] activities are wasteful, lean systems engineering should be the norm [in the delivery of LIPs]." (Kasser, 2013)

In applying "Lean" principles, engineers might progressively "*eliminate waste*" in the design and construction stages of large and complex infrastructure projects. Incidentally, the industry will achieve another important objective, i.e., minimising raw material utilisation as well as delivering cost-effective infrastructure projects.

CHAPTER 6

ENGINEERING PROCESS OUTPUTS

The Cambridge Dictionary (1995) defines a process as, "*A series of actions that you take in order to achieve a result*", which is "the general deliverable of projects, namely assets and supporting documentation so that someone else can learn how to operate the hardware and make the widgets" (Krauss, 2014). It is therefore expected of the processes pertaining to EDD to produce "*results*" in the form of assets/facilities, artefacts, documentation and data.

An attempt is made in the subsequent sections to list the commonly accepted outputs of the engineering processes as discussed in the preceding chapters. By systematically listing and/or discussing these generic outputs, these sections provide a more structured exposition on expected outputs, and provide the engineering team a "set of checklists" to serve as a tool for V&V (i.e., against generated outputs) of the effectiveness of the EDD processes.

The author has catalogued those outputs (some are peculiar to LIPs) in the sequence they have appeared in earlier chapters, in line with the engineering effort required as follows:

6.1 Outputs of the Requirements Analysis Process:

The Requirements Analysis process translates the requirements of the stakeholders into measurable and verifiable System Requirements that lend themselves to the development of a Requirements Breakdown Structure (RBS). Further, System Requirements can be subdivided into unambiguous *categories* of requirements (and such categories were defined in Table 2.1):

- Functional Requirements specify what the system or subsystem must be capable of doing. Using a "verb" language/format, they answer performance questions such as *who, what, where, when, how many, which ones*. For instance, the <bridge> must be capable of <crossing the waterway and accommodating> not less than <1,000 cars> per <hour>. These requirements statements encapsulate the *primary* essence of the system, the reason why it shall exist in the first place.

- Interface Requirements pertain to the "interactions" of the system with the external environment or those imposed on the system by the environment, or those that pertain to the "interactions" between the various objects or components (e.g., parts) within the system itself.
- Non-Functional Requirements (– *ilities*, e.g., reliability, durability) identify and quantify the overall performance of the system in its entire life cycle. They are mistakenly called Quality Requirements.
- Preconditions or Limitations are conditions imposed from outside on the system to be developed, imposing limitations on the solution to be designed. These are based on geometry or spatial arrangement, size, volume, weight and price.

6.2 Outputs of the Functional Analysis Process:

During this Functional Analysis process, the required "functions" are transformed into "solution-independent" subsystems—called "behaviour-enablers". Therefore, this exercise should relate to the Behaviours Domain whereby the designed system is considered from the perspective of "how it shall interact" both internally (i.e., between and among its components) and externally (i.e., with the adjacent or surrounding environment). The major outputs produced during the Functional Analysis Process are as follows:

- **Functional Breakdown Structure (FBS)** serves to primarily define "functionality of the system" without identifying any objects (i.e., components). The functions located *lower down* in the structure collectively represent *higher up* functions. Thus, the FBS depicts a "hierarchy of functions" for consideration in the EDD process.
- **System Breakdown Structure (SBS)** represents the decomposition of the system. All the system components to be designed, produced, maintained and decommissioned are displayed in a clear hierarchy. Thus, key to the SBS is an outlining of "Subsystems" (or "Function Enablers") for design considerations.
- **Specifications or Requirements Breakdown Structure (RBS)** displays the relationship between the requirements, as structured according to various levels of details. By linking those levels, the Requirements Traceability Matrix forms an essential tool for fully determining the scope (or system configuration) and the impacts of any changes to it.

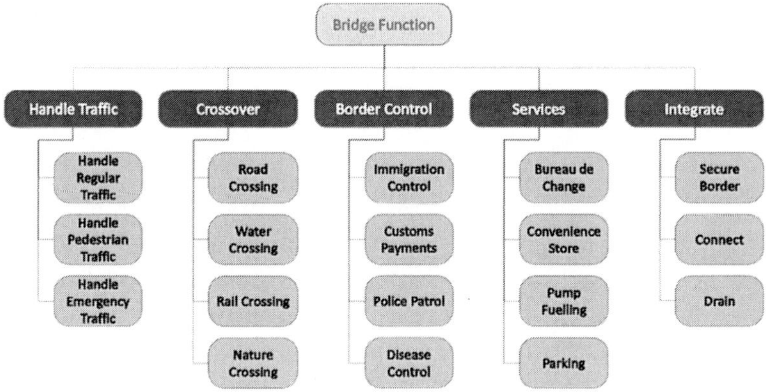

Figure 6.1 – Functional Breakdown Structure (FBS)

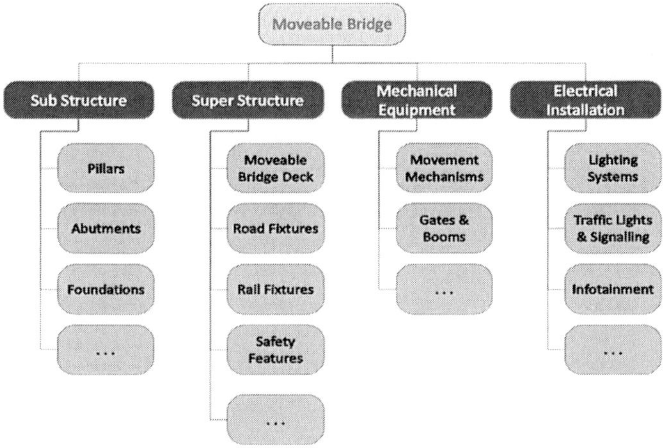

Figure 6.2 – System Breakdown Structure (SBS)

- **Verification Plan(s)** shall be developed for verifying the system based on documented requirements. In prior Requirements Analysis, at the same time as requirements were defined, a Verification Plan (i.e., specifying type of verification to be performed, the criteria, and the applicable tools/methods) was developed so that engineers might consider how to verify that each system requirement has been met. Differences between the end-to-end test and any relevant operational or environmental conditions must be documented in the subsequent Test or Verification Plan (Systems Engineering Handbook, 2007).

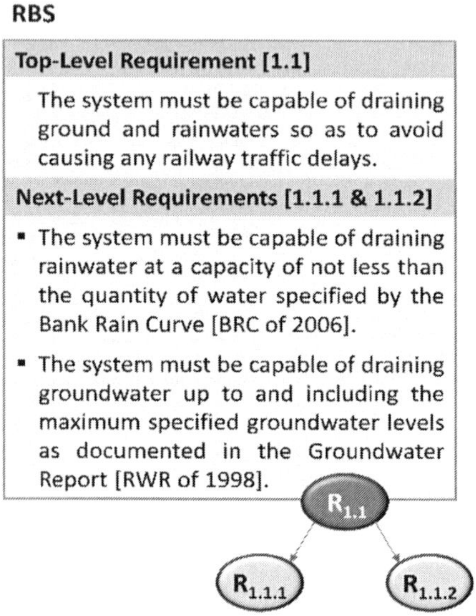

"It is sensible to keep track of traceability of customer requirements to system requirements from the start. This makes it possible to visualize for the client what has been done with his requirements"
— By Alberta Government
(Standards and Specifications Specialist, 2018)

Figure 6.3 – Requirements Breakdown Structure (RBS)
(Adapted: ProRail, 2013)

Further, "The evaluation of system performance also requires the design and construction of a simulation [or test] of the operational environment that realistically challenges the operational system's capabilities" (Kossiakoff et al, 2011).

At this stage, it may begin to emerge that the accuracy of the Functional Analysis process would be heavily dependent on the effectiveness of the Requirements Analysis Process. "The Technical Requirements Definition Process transforms the stakeholder expectations into a definition of the problem and then into a complete set of validated technical requirements expressed as 'shall' statements that can be used for defining a design solution for the Product Breakdown Structure [...] [also called System Breakdown Structure (SBS)]" (Systems Engineering Handbook, 2007), and the same dependency should *logically* spill over to the Design Process.

For that reason, Kossiakoff remarks, "Design must therefore address as a critical constraint the inherent sensing and reaction capabilities of the operator, in addition to a range of associated human–machine interfaces such as the design and placement of [a system's] controls and displays" (Kossiakoff et al, 2011).

6.3 Outputs Produced by the Design Process:

The Design Process puts into effect the "Architecture" and "Measurements" domains (i.e., defining required "components" as well as their shapes, sizes, technical and performance attributes) to produce the actual solutions for solution-independent subsystems. Those subsystems are transformed into specific solution-based designs and, therefore, will produce the products and outputs including the following:

- **Option Specification(s):** Each of the identified options ought to be translated into a set of requirements, which would constitute the System Specifications; NASA shares that perspective and approach when it comes to complex systems. "Once it is understood what the system is to accomplish, it is possible to devise a variety of ways that those goals can be met. Sometimes, that [options identification] comes about as a consequence of considering alternative functional allocations and integrating the available subsystem design options, all of which can have technologies at varying degrees of maturity" (Systems Engineering Handbook, 2007).
- **Variant Specification(s):** The selected options may further spawn a number of technical variants (e.g., concrete versus steel; steel versus fibreglass) as well as the Process Design for the producing of each variant-system. When it comes to large and complex engineering systems, "The growing importance of software-dominated systems, and their inherent complexity and abstractness, has engendered a number of variants of structure" (Kossiakoff et al, 2011).
- **Trade-off (and/or Scoring) Matrices:** "Systems engineering is about trade-offs and compromises, about generalists rather than specialists [engineers]. Systems engineering is about looking at the "big picture" and not only ensuring that they get the design right (meet requirements) but that they get the right design [...] Many trade-off studies [i.e., using trade-off and scoring matrices], typically defined and led by the team systems engineer [i.e., designers], are conducted among subsystem experts immediately as issues occur" (Systems Engineering Handbook, 2007).

  The example provided in Table 6.1 may suggest that for *political* aims the Ferry can still be chosen even though scoring slightly lower. These and similar issues call for an effective collaboration between the engineering team and their colleagues involved in stakeholder management exercise; compromises and trade-offs arise from there.

Design of "Shore-to-Shore Link" – Trade-off Matrix						
Option	*Life cycle costs*	*Safety Requirements*	*Quality of Life*	*Ease of Construction*	*Reliability in Operations*	**Total**
	20%	20%	10%	20%	30%	**100%**
Bridge	7	8	4	9	7	**7.3 √**
Tunnel	4	5	7	3	5	**4.6**
Ferry	5	7	9	7	6	**6.5**

NOTE: Option Scoring ranges from 01 (poor) to 10 (excellent)

Table 6.1 – Example of Trade-off Matrix for "Shore-to-Shore Link"

- **Design Reports, Specifications, Issued for Construction Drawings:** "To determine the degree to which a design will actually result in the desired component performance, and whether or not the component will properly interface with the others, it is necessary to convert its design [through a suitable "how to build"] to a physical entity and to test it" (Kossiakoff et al, 2011).
  This requires that any physical elements be fabricated and individual software components be coded. Prior to system fabrication, reviews are held between the designers and fabrication personnel to assure that what has been designed is within the capabilities of the facility that has to build it (Kossiakoff et al, 2011). All these design activities ought to be documented in structured reports.
  It is also worth noting that the specifications and other Issued for Construction (IFC) Drawings should reflect the Production Process (i.e., the "How to build", not just the scope (i.e., system configurations) of the designed object or system.
- **Verification and Validation (V&V) Reports and Plans:** to address Design (i.e., Reports) and Construction (i.e., Plans) suitability, V&V can form part of the Design Reviews, on a life cycle stage-specific, interim, and final basis. Therefore, NASA maintains, "During Phase B [i.e., Pre-Feasibility, FEL-2], the SE-engine is applied recursively to further mature requirements for all products in the developing product tree, develop ConOps preliminary designs, and perform feasibility analysis of the verification and validation concepts to ensure the designs [of large and complex infrastructure systems] will likely be able to meet their requirements" (Systems Engineering Handbook, 2007).
- **Work Breakdown Structure (WBS):** this artefact describes the structure of related activities (i.e., arranged on the basis of scope, timing or location). In addition, requirements, costs or cost estimates can also be linked to those activities. Moreover, it is often possible

or desirable to group the activities into "work packages", or coherent activities carried out to produce a portion of the system. Furthermore, the "network of work packages" (called *systemigram*) will offer a dynamic *picture* that is useful to engineers and project managers.
The WBS lead to a plan that arranges activities in the three streams of work, namely development, testing and management. This set of interdependent streams of work merge at planned and predefined milestones have activities that implement engineering plans to convert the FCFDS into a practical "*changed situation*" (Kasser, 2015).

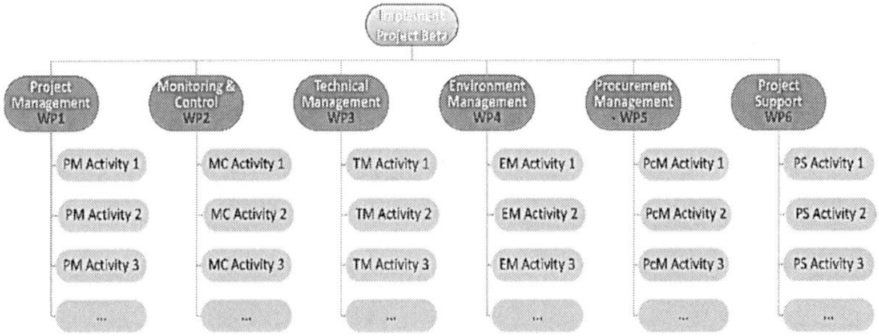

Figure 6.4 – Work Breakdown Structure (WBS)

On the whole, the outputs of the Design Process ought to have reached such a high level of definition at this stage that their accuracy would give assurance and confidence to the decision-makers (e.g., project sponsor, company executives, funders) on the projects to warrant a formal approval to proceed to Execution. In most methodologies applicable to LIPs, this threshold is characterised by a cost/schedule accuracy of -10 to + 15%.

> "Systems engineering deals with problems that generally have a range of equally acceptable solutions [i.e., within tolerance limits]. For example, you are hungry, which is generally an undesirable situation. Your problem is to figure out a way to remedy that undesirable situation by consuming some food to satisfy the hunger." (Kasser, 2015)

Furthermore, the "quasi-irreversible" nature of most construction works is no longer a matter of debate, especially in LIPs. Herein arises the necessity of almost finalising anything related to the "design" before moving to site. Indeed, project managers ought to always bear this stern warning in mind: "*Industry experts are adamant that starting construction without a complete design is always a bad idea*" (This instruction was discussed in Chapter 04).

This prudent attitude ought to equally apply to many other pre-construction activities. In response to the Design, Specifications and IFC Drawings, the contractors shall produce a Method Statement that outlines the hazards and risks involved, and a step-by-step guide on how to get the job done safely.

It details what control measures shall be introduced to ensure the safety of anyone who is affected by the task and/or process. Thus, the contractor should demonstrate how they will provide goods and services in a safe and high-quality manner; this generally involves:

(1) Project details (i.e., site address, description of activities, work dates)
(2) Scope of services to be provided by the contractor
(3) References and technical standards to be applied
(4) Identified scope and site-specific risks and hazards
(5) Proposed delivery method (i.e., work methodology)
(6) Monitoring mechanisms of delivery effectiveness
(7) Quality Assurance, i.e., the core of the Project Execution Plan (PEP)

6.4 Outputs produced by the production process:

In addition to the actual or "physical" production of the system (i.e., new piece of infrastructure and/or improved operational service), the following artefacts or pieces of information are created during this important process:

- **Deployed System:**
  (1) "As-Built" documentation, when applicable
  (2) Verification or Testing Report(s) on both System and Process
  (3) Validation or Acceptance Report(s) on System and Process
  (4) Updated Inspection Plans for post-production stage, e.g., PIRs
  (5) Emergent Properties:
    i. These emergent properties are not necessarily as expected at Design and Development. Since infrastructure are SoS, they usually exhibit emergent behaviours, which requires a more holistic treatment and a somewhat different set of analytical skills (Glass et al, 2008).
    ii. "The SoS performs functions and carries out purposes that are not necessarily associated with any component system. These behaviors are emergent properties of the entire SoS and not the behavior of any component system [alone]" (Kossiakoff et al, 2011)—Figure 6.5 seeks to classify the "desirability" of such behaviours. Systems may exhibit *intended*, flawed/defective and serendipitous behaviours.

- **Decommissioning/Disposal Plans:**
  These plans concern themselves with the steps and resources involved in removing any deployed system at the end of its cycle, including the by-products and elements disused during operations. Reverse-Logistics approaches and plans should be progressively elaborated to allow for the timely removal of such *disposable* items.
- **Operations and Maintenance Plans:**
  Forsberg (2005) maintains that, "All operations and maintenance staff should be in place and properly trained. The maintenance plans for the system should be reviewed […] to make sure that all maintenance procedures and hardware and/or software maintenance records are in place and adequate to properly maintain the system" (Forsberg et al, 2005). Hence, Maintenance Plans are developed and put in effect in due course to sustain high "system operability".

	Desired	Undesired
Known	Purpose of system	Flaw/Risk to be mitigated*
Unknown	Serendipitous	Defect

- Flaw occurs every time, e.g. design flaw
- Risk has a probability of occurrence
- Serendipitous also provides opportunities

Figure 6.5 – Typology of Emergent Properties (Adapted: Kasser, 2016)

Due validation of the "constructed systems" with relevant stakeholders (e.g., users, client) might constitute the ultimate step of the System Design and Development process. Unless V&V had duly taken place between Specifications and Design, and between Design and the planned System, conducting it at the tail-end might prove futile, as Figure 6.6 suggests.

In their totality, the listing and succinct discussions pertaining to EDD shall provide those engineers involved in large and complex infrastructure with a canvas to guide them as to the effectiveness of EDD processes by way of gauging major outputs. Understanding that the broader scope of the outputs constitutes the totality of deliverables, the listing of the engineering outputs will assist engineering and project management teams in determining the "*Tasks, resources and risks*" involved in carrying out the engineering endeavour.

With the above considerations in mind, one of the critical "deliverables" of the Design and Development Processes could be in the form of "Technical Feasibility" document that determines to what extent a solution exists (e.g., a new deep-sea port) that adequately solves the problem (e.g., the safe berthing cargo ships) that confronted a particular organisation or society.

Moreover, infrastructure projects being socio-economic investments, the said "Technical Feasibility" document ought to also include an estimated budget (or financial rendition of "Tasks, recourses, and risks"), as well as delivery timelines. Kossiakoff concedes that, "It requires the investment of time and effort to gain experience and an extensive broadening of the engineering base" (Kossiakoff et al, 2011).

Accordingly, adequate resources are needed to support engineering efforts:

> "Determining resources necessary to accomplish the objectives […] The product produced by these [planning] activities is the first draft of the plan. This draft is an ideal plan that does not take into account any limitations of resources such as budget, personnel and time." (Kasser, 2015)
>
> "The draft work packages [relating to engineering efforts] are adjusted to meet the time and cost budgets. If the resources available [i.e., personnel, funds and time] do not meet the draft budgets, then low priority work packages may have to be flagged, removed or delayed." (Kasser, 2015)

Engineering teams involved in LIPs should not forget to include the scope and costs of Operational Readiness, which is all about "*making the organisation ready to safely receive and effectively utilise the 'products of projects' (which are systems)*" (Mabelo, 2020b). Citing various sources (such as INCOSE, 2011; GP Strategies, 2016), the author has discussed this concept in one of his previous publications (Mabelo, 2020b) as follows:

> "Ultimately, the Transition Process transfers custody of the system and responsibility for system support from one organizational entity to another. This includes, but is not limited to, transfer of custody from the development [i.e., engineering] team to the organizations that will subsequently operate and support the system. Successful conclusion of the Transition Process typically marks the beginning of the Utilization Stage of the system-of-interest."
>
> "While this is a relatively short process, it should be carefully planned to avoid surprises and recrimination on either side of the agreement. Additionally, transition plans should be tracked and monitored to ensure all [readiness-related] activities [and artefacts] are completed to both parties' satisfaction, including resolution of any issues arising during transition."

> "In fact, GP Strategies (a global company based in the USA) maintains that, a key goal of capital projects is to achieve a State of Operational Readiness in order to achieve successful Start-Up on time and within budget […] Another goal is to reach a State of Operational Excellence [in terms of desired performance objectives] as soon as possible after Start-Up."

Considering that systems are indeed designed, constructed and deployed for the sake of improving (or establishing, *in the first place*) effective and sustained operations (Mabelo, 2020b), the following operational objectives and parameters must be assessed to confirm the suitability of the new infrastructure system, as deployed in its intended environment:

<table>
<tr><th>Operational Performance Objectives</th><th>EDD Implications</th></tr>
<tr><td>Quality:
Conformance to specifications/requirements</td><td rowspan="6">Have EDD outputs resulted in a system that duly and fully meets most criteria of Quality, Speed, Dependability, Flexibility, Cost and Sustainability in operations?</td></tr>
<tr><td>Speed:
Producing goods or services fast, minimizing the time between a customer orders and delivery</td></tr>
<tr><td>Dependability:
Delivery on time, keeping to delivery promises (e.g., which is contingent to regular maintenance)</td></tr>
<tr><td>Flexibility:
Being able to change far enough and fast enough to meet customer requirements</td></tr>
<tr><td>Cost:
Producing goods and services at a cost which may be priced appropriately for the market</td></tr>
<tr><td>Sustainability:
Producing goods and services that minimise any "adverse effects" on the environment and users</td></tr>
</table>

Operational Readiness Parameters	EDD Implications
Legal and Statutory Readiness: *Securing licenses to operate or applying for new (by)-laws to allow or protect operations*	**Have EDD outputs resulted in a system that fully satisfies *Legal & Statutory (regulator's approval), Human Capital (availability of O&M skills), Processes/Support, Bulk Services and/or Infrastructure (existing and planned capacity)* requirements?**
Human Capital Readiness: *Addressing skills and capacity shortages or any other organisational design flaws (e.g., structure)*	
Processes/Supporting Systems Readiness: *Ensuring workflows and related management systems (i.e., ICT platforms, ERP, etc.) are able to support changes in operating environment*	
Bulk Services/Infrastructure Readiness: *Availability of bulk services/infrastructure and supplies to support/sustain operations*	
Commercial Agreement: *Signing Offtake Agreements before start-up (Assuming availability of willing customers and existence of workable, effective trading platforms)*	**Does the EDD produced system conform/align to *market demand* (client, contract/trading policy)?**

Table 6.2 – Operational Performance/Readiness and EDD Implications

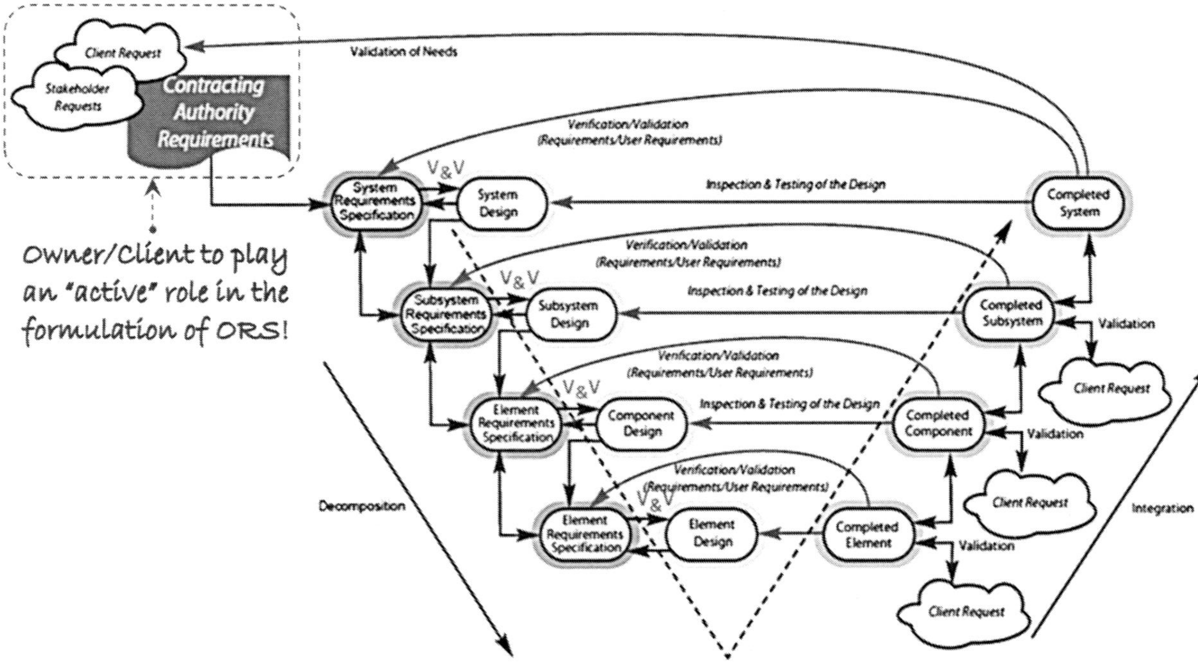

Figure 6.6 – Vee Model Showing Design and Development (as per ProRail, 2013)

CHAPTER 7

THE ENGINEERING PROCESS AND PROJECT LIFE CYCLE

When it comes to infrastructure projects, complexity (e.g., in integration) is the order of the day. Bar-Yam argues: "A fundamental reason for the difficulties with modern large engineering projects [i.e., LIPs] is their inherent complexity. Complexity is generally a [salient] characteristic of large engineering projects today" (Bar-Yam, 2014).

The increasing complexity of infrastructure projects has created the need for design professionals trained in all phases of the project life cycle and for them to develop an appreciation of the *building* as an advanced technological "system" that requires close integration of many subsystems and their individual elements including sustainability (Hussin and Omran, 2009). Therefore, project life cycle shall take a literal *centre-stage* (Mabelo, 2016).

In discussing the concept of project life cycle model, Archibald similarly maintains that,

> "A well-documented Project Life Cycle Model enables us to apply Systems Thinking to creating, planning, scheduling, and managing the project through all of its phases, and to evaluating both the success and the value of both the project and results [e.g., design and development outputs] that the project has produced" (Archibald, 2012).

More significantly, Archibald further argues, "Unless a well-documented, integrated, understandable picture of the overall life cycle process—the model—for each project category/sub-category exists, it will be difficult to achieve the full benefits of modern, systematic project management [as applied to large infrastructures]" (Archibald, 2012).

SE focuses on the needs of the users and other stakeholders throughout the life cycle "to enable the realisation of successful systems" (Fossnes and Forsberg, 2006); from this perspective, a life cycle model is needed to guide the engineering effort as follows:

I. To understand how the evolution of the system life cycle relates to the SE process, it is useful to develop a model of its structure down to the second level of subdivision (i.e., Concept Definition). The Concept Development stage involves three phases: Needs Analysis, Concept Exploration, and Concept Definition (see Figure 2.3 in Chapter 02).

II. Figure 7.1 below represents the activities, inputs and outputs of the constituent phases of the Engineering Development stage in the system life cycle—e.g., the Advanced Development, Engineering Design and Integration and Evaluation, starting from the Defined System Concepts.

Furthermore, specifically for large and complex infrastructure projects (LIPs), the US-developed Intelligent Transportation Systems (ITS) Model exhibits how the System Design and Development process could (or rather should) match the *generic* system life cycle. For instance, the model links System Requirements to Preliminary Engineering, while locating System Validation (V&V) at the end of Construction.

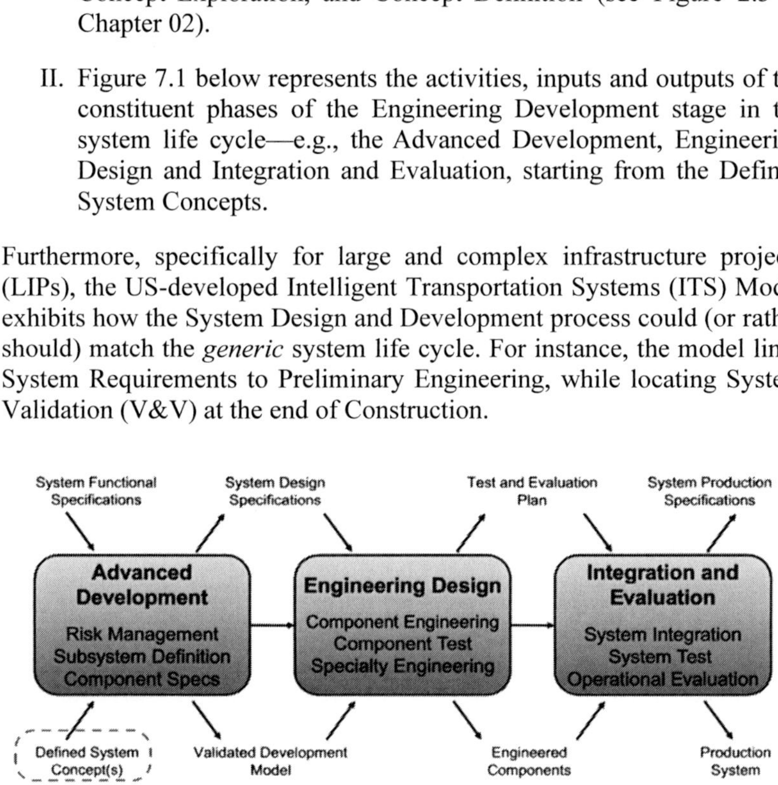

Figure 7.1 – Engineering Concept and Specifications (as per Kossiakoff, 2011)

Figure 7.2 below would combine a "waterfall" project life cycle with the Vee Model applicable to System Design and Development (as discussed in Chapter 04). This approach is in line with SE principles and concepts; effective LIPs Delivery should incorporate both Waterfall and Vee life cycle models.

Moreover, a close scrutiny of Figure 7.1 reveals that "system specification" should contain all the information that defines the system, evolving from the Functional to Design to Production specifications. A significant SE tenet consists of keeping design/solution separated from requirements/needs as much as possible to keep the solution space clear. As Figure 7.2 shows, requirements and design should develop in parallel and/or iteratively to system development, while maintaining that, "*(3) 'a structural separation of requirements and solutions [is] embedded in the life cycle methodology*".

Moreover, project teams are nowadays using Agile approaches in a variety of industries (e.g., manufacturing, finance)—beyond software development (ICT) where it originated. "Today, [most] project leaders and teams find themselves in an environment disrupted by exponential advances in [pioneering] technology and demands from customers for more immediate delivery of value [...] In addition, the first principle of Agile places customer satisfaction as the highest priority and is key in delivering products and [or] services that delight customers [...] Rapid and transparent customer feedback loops are readily available [...] More mature organisations are increasingly prone to being highly complex and potentially slow to innovate, and lag behind in delivering new solutions to [...] customers." (Project Management Institute, 2017)

This too shall tell the sad story of entities involved in delivering large and complex infrastructures; not only do they grapple with *delighting* the customer (i.e., users being expected to enjoy whatever infrastructure they are provided with), but they also lag behind in bringing about new solutions. The modern tar-road, for instance, has remained the same technically and conceptually (in both "What to build and "How to build") for decades now.

"New design, problem-solving, and not-done-before work [i.e., infrastructure based on novel, innovative ideas] is exploratory" and, thus, entails Agile approaches (Project Management Institute, 2017). This could apply to any innovatively designed infrastructure system, whether road/rail networks, water and power plants, sea/airports, hospitals or mega-malls.

Since Agile life cycle methodologies leverage on aspects of "iterative" and "incremental" characteristics, design teams using Agile approaches iterate over the product design to create finished deliverables [i.e., assemblies, subsystems]. Agile teams gain feedback and provide customer visibility, confidence, and control of the product. Thus, design teams are able to release earlier, and the project provides an earlier return on investment by executing the highest-value work first (Project Management Institute, 2017).

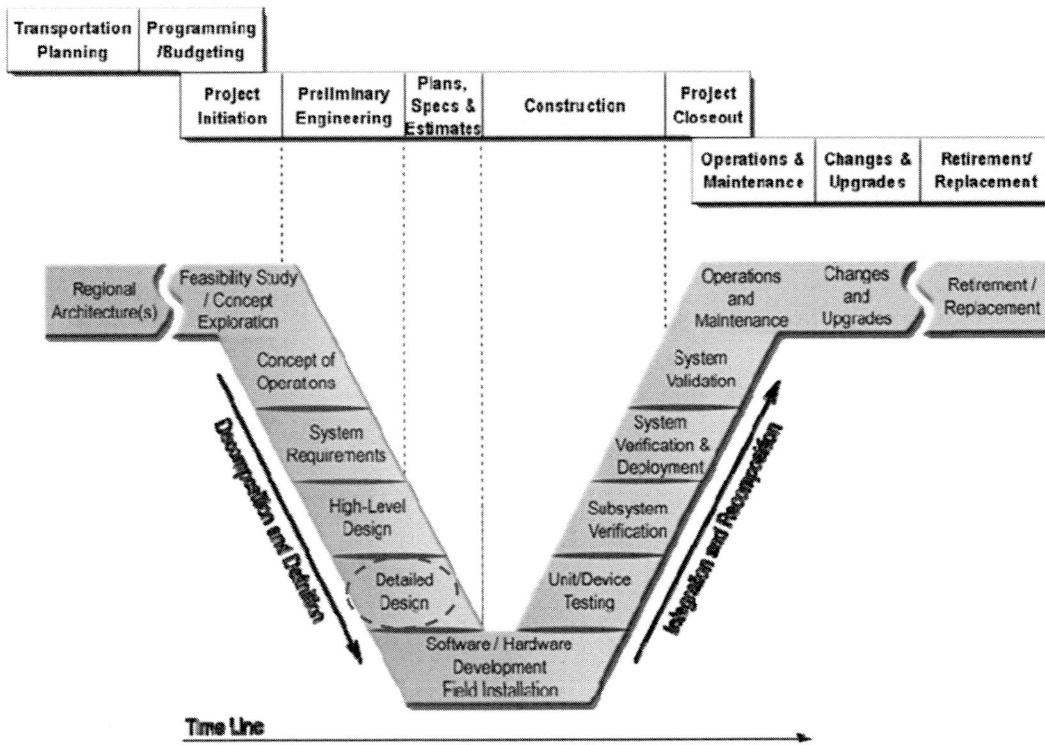

Figure 7.2 – ITS Model Combining Vee and Waterfall (as per Ryen, 2008)

Figure 7.2 suggests that a suitable System Design and Development life cycle combines a "waterfall" with the Vee Model and that such a life cycle accommodates a V&V exercise (i.e., customer feedback) between each two sequential phases. It is not far-fetched to solicit such feedback including within each and every phase. This applies to the Planning Phase (i.e., Front-End-Loading, FEL phases) and herein lies a good place to become "agile". Indeed, Construction works (e.g., foundation piling, pouring of concrete) may not practically accommodate Agile as "investment decisions may be *irreversible* once production and assembly of the product begins" (Whelton and Ballard, 2002).

"For complex systems the iterative process of specifying is repeated at several levels of detail. Design choices are made based on the existing set of requirements. Requirements can be derived from this design, resulting in design choices at a more concrete and more specific level (or a finished product). This iterative process [which only covers the design part] results in a breakdown of the system to be realised. The result of completing this iterative process during the development phase is a specified system, including the design that has been verified accordingly" (Standards and Specifications Specialist, 2018).

To deliver the final product (i.e., the new system), "The designed system is then realised from the bottom up and mainly verified using the design. The specification level of detail is determined by the risk dossier, the complexity of the system to be realised and the required [detailed] level of concreteness of information to be able to realise the intended system [i.e., large and complex infrastructure]" (Standards and Specifications Specialist, 2018).

It transpires that from the Requirements Analysis stage, the Engineering Design shall *steadily* (and somewhat iteratively) progress into Detailed Design over the project life cycle. Figure 7.3 below depicts how the project undertaking should shift from the "Learning Realm" (i.e., with minimal design efforts—and emphasis on commercial aspects) to the "Optimisation Realm" (with intensified design efforts).

Not many project practitioners, including those involved in large and complex projects, might fully appreciate the practical implications of this life cycle map. In fact, the author has come across a multibillion Rand project that failed to *first* explore and address the business considerations through the "Learning Realm" (Conceptual and Pre-Feasibility).

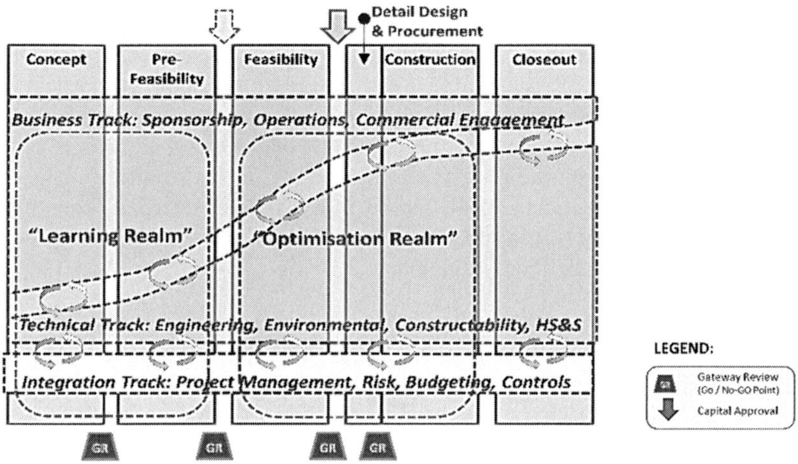

Figure 7.3 – Business vs. Technical Emphasis in Project Life Cycle (Adapted: Mabelo, 2016)

The project board had incorrectly structured this mega project as a "programme" with the aim of proceeding with a separate "engineering project" and, therefore, denied themselves the opportunity to investigate alternative designs. They evolved headlong with a single engineering option to produce detailed designs and cost estimates for most of its elements.

Although the organisation had no previous experience in projects of such a magnitude or complexity, they had satisfactorily delivered at least two projects of a similar scope; the project team should have learned from those undertakings. For instance, the performance of container terminals is measured by its throughput (i.e., volume of twenty-foot container equivalent unit (TEU) per annum), with the ratio of such a throughput divided by the capital outlay and operating costs largely informing the minimum tariff required from prospective customers.

The abovementioned TEU ratio is a significant commercial consideration in gauging the business viability of a container terminal. Should a candidate project fail this acid-test, chances are that the terminal might not prove economically feasible/viable. In the case of this mega project (new container terminal), assuming operating costs would be roughly the same, the two prior container terminals had a ratio of respectively 1.04 (1.0 billion Rand for 0.96 million TEUs p.a.) and 1.20 (4.5 billion Rand for 3.75 million TEUs p.a.) compared to an outlandish ratio of 55.56 (500 billion Rand for 9 million TEUs p.a.) for the proposed container terminal.

Surely no expert-facilitated workshops or overseas trips were needed to realise that the proposed megaproject will probably prove non-economically viable, merely on the basis of its capital or TEU ratio being so massively out of range. The commercial implication, by a rough estimate, would translate into a tariff of ± 50 times higher than what the company might have been charging its current clients. The author was diligent from the outset in pointing out this predicamental concern to the project team; at their own peril, they dismissed his advice until two years and a few hundred million Rand later when a dead-end prevailed.

Indeed, in the early phases (Concept and Pre-Feasibility), most of the emphasis must be on "Business" aspects (e.g., policy/strategy alignment, commercial validation) instead of "Technical" aspects (i.e., engineering details) and vice-versa from the Feasibility Phase (FEL-3). Detailed Design, however, remains one of the essential activities at Feasibility.

> "Detailed Design involves the process of successively breaking down, analyzing, and designing the structure and its components so that it complies with the recognized codes and standards of safety, and performance while rendering the design in the form of drawings and specifications that will tell the contractors exactly how to build the facility to meet the owner's need. During this phase, detail[ed] design of the work, contract documents, detail[ed] plan, budget, estimated cash flow, regulatory approval, and tender/bidding documents are prepared. Depending on the type of contract the owner would like to have for completing the facility, the designer (i.e., Consultant) can start preparing the Detailed Design." (Rumane, 2018)

Detailed Design is in fact crucial to project delivery. In fact, Rumane recommends that, "The success of a project is highly correlated with the quality and depth of the engineering plans prepared during this phase [...] Detail[ed] Design activities are similar, although more in-depth than the design activities in the preliminary design stage. The size, shape, levels, performance characteristics, and technical details and requirements of all the individual components are established and integrated into the design. Design engineers of different trades have to take into consideration all these as a minimum while preparing the scope of [the] works" (Rumane, 2018).

Accordingly, the SE-based Project Life cycle Methodology or SE-PLM model (Mabelo, 2016) features the System Domains, the Life cycle, the Engineering Process, and the Operational Environment. It also indicates the essential Post-Implementation Review (PIR)—which is usually omitted, including in LIPs where effective transition to operations should be tested.

The author upholds the necessity and requirements of the "holistic" system life cycle. "It transpires from the above that the life cycle model constitutes the basis of any project delivery methodology; therefore, addressing shortcoming and inadequacies in the delivery of LIPs should start by applying enhancements to the project life cycle itself, as the 'racetrack' of the project delivery efforts. This study is thus justified in envisaging an enhanced, SE-based [project life cycle] model." (Mabelo, 2016)

The SE-PLM model is illustrated in Figure 7.4 below. The Primary Scheme (from FEL-1 to Closeout) should be repeated (in successive layers) to cater for each entity-level of the System (i.e., system, subsystems, assemblies, and elements). The Vee Scheme (below it) depicts SE processes that deal with the technical aspects of the project life cycle, including the engineering process of Detailed Design, even with the opportunity to apply Agile approaches during the FEL phases—since Construction will not allow it.

The Primary Scheme concerns itself with the progression of the Realisation-System (i.e., teams, processes, etc., put in place to deliver the products of the project) through the phases of the project life cycle (Conceptual, Pre-Feasibility, Feasibility and Execution). This is where the Integration Track interfaces with the Business and Technical Tracks, for instance, in managing the engineering endeavour. Thus, the matching Vee Scheme concerns itself with the technical development of the Solution System (i.e., product-oriented processes). In line with the Vee Model, this Vee Scheme evolves through the Decomposition and Production of the "components" (according to the System Hierarchy) of the Solution System in order to expedite the creation of the Solution System as envisaged.

Further, the rightward sphere represents the Context-System (Operational Environment, where the Solution shall be deployed, where it shall live). This domain encapsulates three sub-domains, namely, the (1) Sustaining-Systems (i.e., systems needed to sustain the Deployed System), (2) the Collaborating-Systems (i.e., systems that are to collaborate with the Deployed System), and (3) the Competing-Systems (i.e., systems vying for resources and/or space with the Deployed System). It is crucial that the design and development of systems (i.e., product of projects) takes into due account the needs and requirements of every sub-domain of the Context-System. Unless this is achieved, the new system might not perform as intended; the Tacoma Narrows Bridge had definitely failed in that regard.

Chapter 05 discussed the necessity of designing LIPs as CAS in order to avoid *unintended* dynamics as a result of interactions with the environment:

> "Systems Thinking is especially useful for defining problems [before proposing solutions] [...] By shedding light on the root causes of a problem and letting you anticipate the multiple consequences of your solutions [e.g., unintended behaviours], it can help you avoid solutions that only spawn more difficulties [in operations]." (Mabelo, 2020; citing Goodman, 1997)

The SE-PLM model, indeed, puts a strong emphasis on the Operational Environment. The author has already submitted that the primary purpose of any project (including large and complex infrastructures) should entail the establishment and/or improvement of the Operational Environment where the newly engineered "solution system" is intended to be deployed (Mabelo, 2020b). This view is consistent with Systems Engineering, as Kossiakoff maintains regarding the "holistic" approach to project design and delivery:

> "Systems engineering differs from traditional disciplines in that (i) it is focused on the system as a whole; (ii) it is concerned with customer needs and operational environment; (iii) it leads system conceptual design; and (iv) it bridges traditional engineering disciplines and gaps between specialties. Moreover, systems engineering is an integral part of project management in that it plans and guides the engineering effort." (Kossiakoff et al, 2011)

Furthermore, Ryen discusses the governance provisions of the Vee Model; "The 'V' diagram [as fused in SE-PLM] is similarly punctuated by a series of major milestones [...] where the output of the previous step is reviewed and the customer and project team determine whether the project is ready [i.e., feasible, viable] to move to the next step in the process" (Ryen, 2008).

It might therefore prove imprudent to wait for the completion of the whole system before verifying that requirements have been met and that the project has been largely successful. These key provisions are achieved through two separate governance processes as follows:

(1) Gateway Reviews usually take place between every two-consecutive project life cycle phases as indicated in Figure 7.3 above. The essence of Gateway Reviews, which involve the three streams (i.e., commercial, technical, and integration) being:
   - (a) *Backward-looking Review*: it seeks to establish to what extent the works of the previous/ending phase were executed in compliance with the relevant portions of the approved project life cycle methodology and/or regulatory frameworks
   - (b) *Forward-looking Review*: it seeks to establish the degree of commercial and technical viability (as per the outcome of the works of the previous phase) and the readiness to proceed to

the next phase, to support Go/No-Go decisions, and/or justify investment decisions regarding the ensuing phases of the project.

(2) Technical Reviews, as a purview of the engineering teams, shall ascertain that the right project/system requirements are being satisfied at specific life cycle milestones:

(a) OCR (Operations Concept Review): seeks to establish to what extent Needs Identification has yielded the right (set of) Concept of Operations (ConOps)

(b) SRR (Systems Requirements Review): seeks to establish to what extent Requirements Formulation has delivered the right Systems Requirements, which ought to serve as input to the subsequent System Design process

(c) PDR (Preliminary Design Review): seeks to establish/confirm to what extent Preliminary Design has delivered the right design to serve as input to the ensuing Critical Design sub-state of the System Design State

(d) CDR (Critical Design Review): seeks to establish/confirm to what extent the System Design is ready to commence the Subsystem Construction/Fabrication

(e) TRR (Test Readiness Review): seeks to establish to what extent the Subsystem Construction is ready to sustain Subsystem-Testing (and Validation, is needed)

(f) IRR (Integration Readiness Review): seeks to signal/confirm termination of the Subsystem-Testing State and the commencing the System Integration State

(g) DRR (Delivery Readiness Review): seeks to establish to what extent System Integration and System Test are successful in terms of commencing activities that deliver the system, leading to terminating (i.e., Close-Out) the project.

A broader (albeit non-exhaustive) list of Technical Reviews is included below in Table 7.1; it seeks to describe in further details the various technical reviews as well as the appropriate point(s) in the system life cycle where they should apply.

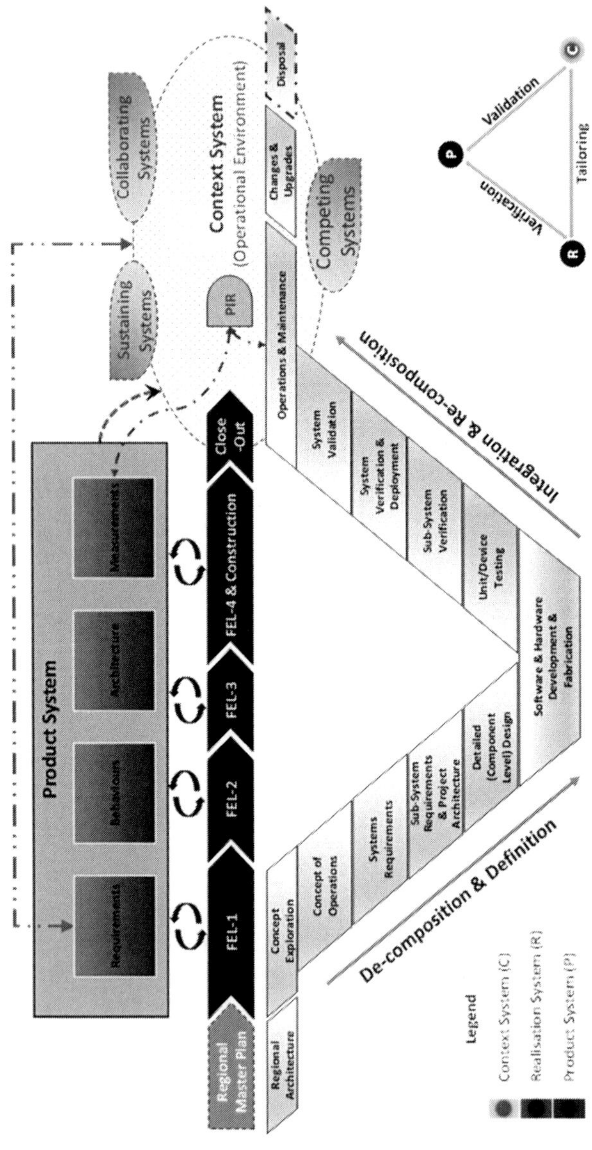

Figure 7.4 – Systems Engineering-based Project Life Cycle (as per Mabelo, 2016)

Review	Description (as per SEBoK)
Alternative Systems Review	A multi-disciplined review to ensure the resulting set of requirements agrees with the customers' needs and expectations.
Critical Design Review (CDR)	A multi-disciplined review establishing the initial product baseline to ensure that the system under review has a reasonable expectation of satisfying the requirements of the Capability Development Document within the currently allocated budget and schedule.
Functional Configuration Audit	Formal examination of the tested characteristics of a configuration item (hardware and software) with the objective of verifying that actual performance complies with design and interface requirements in the functional baseline.
In-Service Review	A multi-disciplined product and process assessment to ensure that the system under review is operationally employed with well-understood and managed risk.
Initial Technical Review	A multi-disciplined review to support a programme's initial Programme Objective Memorandum submission.
Integrated Baseline Review	A joint assessment conducted by the government programme manager and the contractor to establish the Performance Measurement Baseline.
Operational Test Readiness Review	A multi-disciplined product and process assessment to ensure that the system can proceed into Initial Operational Test and Evaluation with a high probability of success, and that the system is effective and suitable for service introduction.
Production Readiness Review (PRR)	Examines a programme to determine if the design is ready for production and if the prime contractor and major subcontractors have accomplished adequate production planning without incurring unacceptable risks that will breach thresholds of schedule, performance, cost, or other established criteria.
Preliminary Design Review (PDR)	A technical assessment establishing the physically allocated baseline to ensure that the system under review has a reasonable expectation of being judged operationally effective and suitable.
System Functional Review (SFR)	A multi-disciplined review to ensure that the system's functional baseline is established and has a reasonable expectation of satisfying the requirements of the Initial Capabilities Document or draft Capability Development Document within the currently allocated budget and schedule.
System Requirements Review (SRR)	A multi-disciplined review to ensure that the system under review can proceed into initial systems development, and that all system requirements and performance requirements derived from the Initial Capabilities Document or draft Capability Development Document are defined and testable, and are consistent with cost, schedule, risk, technology readiness, and other system constraints.

System Verification Review (SVR)	A multi-disciplined product and process assessment to ensure the system under review can proceed into Low-Rate Initial Production and full-rate production within cost (programme budget), schedule (programme schedule), risk, and other system constraints.
Technology Readiness Assessment	A systematic, metrics-based process that assesses the maturity of critical technology elements, including sustainment drivers.
Test Readiness Review (TRR)	A multi-disciplined review designed to ensure that the subsystem or system under review is ready to proceed to a formal test.

Table 7.1 – Technical Reviews and Life Cycle Milestones (Source: SEBoK, 2011)

Further, the term "Operational Environment" as used here includes not only the external physical conditions under which a system operates, but also any other factors such as the characteristics of all system interfaces, procedures for achieving various levels of system operational readiness, factors affecting human–machine operations, logistic issues, and maintenance and so on.

> "The role of systems engineering is essential throughout, but perhaps more so in the early phases [i.e., where "*grasping the soul of the system*" occurs], where an understanding of the operational environment and the availability and maturity of applicable technology are most critical. In later phases, environmental, interface, and other inter-element requirements are the special province of systems engineering." (Kossiakoff et al, 2011)

Thus, failure to apply SE in the early phases can (and will) yield deficiencies downstream. A team of chemical engineering consultants has recently discussed a notable case of design inadvertence with the author. The notion of "engineering/design inadvertence" is "*The absence of attention or care; failure to carefully and prudently follow due process*". This concept is borrowed from the Legal Dictionary where "inadvertence" is defined as:

> "*The absence of attention or care; the failure of an individual to carefully and prudently observe the progress of a court proceeding that might have an effect upon his or her rights*".

This instance of engineering inadvertence pertained to an under-performing process plant at a mining operation, a "concentrator" where valuable minerals (i.e., concentrates) were liberated from invaluable gangue (i.e., tailings). However, something not so major had made this run-off-the-mill mining operations turn into a *peculiar* situation in the sight of management.

Although the mine and its attendant process plant has been operating for years, there came a time where the mine's executive management grew extremely dissatisfied with their performance, particularly over the low rate of metal recovery at the process plant. They then appointed a consulting team to investigate the situation and explore possible ways of addressing those poor throughput issues. After all, profitability was at stake here!

Having conducted a desktop review of the processing method to understand the operations of the plant (i.e., processes, machinery, and instrumentations), the team scheduled a short site visit with the aim of instigating a "metal accounting audit" to identify the stages/points where suspicious loss of metal might have been occurring. The appointed engineering team, as a matter of course, focused their attention on the interfaces where "exchanges of material, energy and data" were supposed to take place—in line with Systems Thinking.

Great was their disappointment when the plant manager acknowledged the problem, but irritably conceded "*there was nothing he could do about it*". There was basically no way anyone could attempt to enhance throughput on the plant since its initial (i.e., the "As-designed") configuration made no accommodation for gauging/measuring of material input and output at process interfaces, e.g., before and after every "flotation-cell" to determine to what extent the loss of mass and/or concentration might be occurring at that particular system-entity.

Unless and until some form of measurement devices were installed at interfaces, a "metal accounting audit" was rendered impractical and, thus, investigating possible causes or triggers of mass and concentration losses could not proceed. The real trouble was that retrofitting such measuring apparatuses at those interfaces would have necessitated a "redesign" of the whole plant to allow not only for the physical space for their fitting, but also to cater for their bearing and possible impacts on the performance of the entire process flow.

It would appear that the engineering team that was initially appointed to design the chemical plant had failed to follow the practical guidance from Oehmen regarding system components and interfaces:

> "Modules are self-contained units [i.e., subsystems, assemblies] complete with interfaces that conform to the plug-and-play passive infrastructure. They can [and should] be dragged and dropped into a system of response capability with relationships to other modules connected through the passive infrastructure, and not connected directly module-to-module. Modules are

> encapsulated so that [adequately flexible] interfaces conform to the passive infrastructure, but their methods of functionality are opaque to other modules. New modules can be added to module pools and new pools of modules can be added asynchronously." (Oehmen, 2012)

As usual, upstream design errors manifested themselves downstream, "*in later life cycle phases*"—with the implications of an overhaul shutdown (thus, loss of revenue) and a costly retrofitting of modules before any performance improvements could take place.

> "Downstream troubles, however, are symptoms of upstream shortcomings; indeed, Oehmen (2012, p. 70) maintains, '*Upstream activities must be held responsible for issues they cause in downstream activities*'. Construction and operations problems might need a recourse upstream back to planning and development (in order for such to be addressed in an integrated manner). Attempting to quick-fix a *seemingly isolated* faulty element of the system may have negative impact on other elements of the system, not necessarily in close proximity of time or location [...] — at times making the problem worse." (Mabelo, 2016)

Hence, the author would crystalize this valuable lesson-learned in Systems Engineering jargon as follows:

> "What the systems engineer shall or should not do at any point in the system life cycle will depend on what was or was not done upstream on the system and will determine what the eventual system might or might not do—for instance, its architecture (i.e., subsystems, asse) will depend on the prior Functional Allocation—and will determine and/or govern the eventual system 'behaviours' (as well as the system's overall performance)."

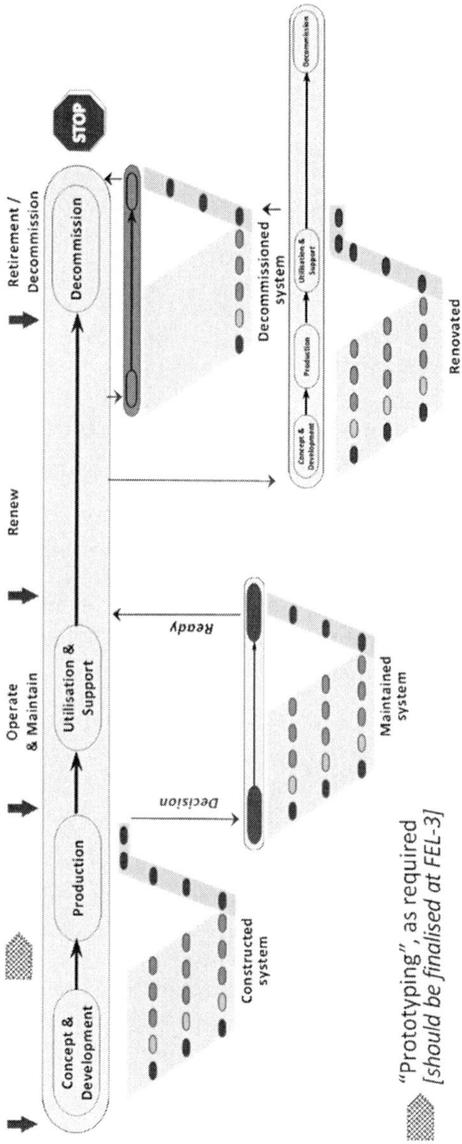

Figure 7.5 – Extended System Life Cycle (as per ProRail, 2013)

CHAPTER 8

FAILURE MODE AND EFFECTS ANALYSIS PROCESS

According to INCOSE, the Concept of Operations (ConOps) describes the intended operation of the system to be developed and helps the systems engineer to understand the context within which its requirements need to be captured and defined (Fossnes and Forsberg, 2006). It follows that the Engineering Design and Development (EDD) process should basically aim to design a "system" that operates in the intended manner.

Further, a point has been made from the onset that, "Not all that engineers build has become successful; occasionally, catastrophic failures occur" (Haik and Shahin, 2011). Thus, the primary aim of the EDD process is not only about designing the Solution System (and its construction/manufacturing process), but it is also about detecting errors and averting failures that might occur in operations.

From this perspective, Failure Modes and Effects Analysis (FMEA) is both a product and process improvement tool (i.e., Quality Management) and a risk mitigation technique (i.e., Risk Management) conducted by specialty (e.g., safety, reliability) engineers and other practitioners, although FMEA, as a bottom-up reliability analysis method, is not *sufficient* for safety.

For instance, a particular fittings, product, machine, or structure might fail physically as a result of a faulty part and/or user behaviour. And conversely, a manufacturing or business/administrative process might fail operationally due to incompetent personnel, flawed design, or faulty equipment/controls. Designers being human, it will be expected that such "inadvertences" will creep into the engineering process to compromise and bedevil the system.

Therefore, FMEA is a suitable method of investigation for determining how a product, process, and/or system might fail and the likely effects of particular modes of failure (in line with the selected ConOps)—FMEA can be used in EDD to help engineers as follows:

i. Assess the possible ways (i.e., modes) in which failures might occur
ii. Assess the magnitude of the effects of failure
iii. Uncover the possible cause(s) of failure
iv. Understand what can be done to prevent or mitigate the prospect of such failures.

"Failure modes" are of interest to operations and construction/assembly engineers, whereas "causes" (i.e., design and process weaknesses) are of interest to design engineers. Consequently, preventing engineering failures will entail system and process design revisions (Biswas, 2015).

By convention, this is achieved by investigating the possible combinations of "effect–cause–detection" and by numerically rating each of the following three elements: effects, causes and detection in the context of the system's susceptibility to failure (Snee and Rodebaugh, 2020).

While both academics and practitioners would agree with the above purpose of FMEA, it shall be noted that listing each element as a separate step might create a false impression of "linearity", even ignoring that failure elements (i.e., modes, causes, effects) are interconnected and interact with each other throughout the system life cycle from design to construction to operations—which is consistent with Systems Thinking. Securing reliable components will not necessarily assure reliability/safety of the whole, e.g., a plant that operates *reliably* but still emits toxins into the atmosphere is not safe.

Carlson (2012) provides a more *systemic* rendition of the FMEA purposes:

(i) To identify and fully understand potential failure modes, their causes and their effects on the system or end-users (i.e., Systems Thinking), for a product/process
(ii) To assess the risk (i.e., to the system, by the system) associated with the identified failure modes, effects and causes, and the prioritise issues for corrective actions
(iii) To identify and effect corrective actions to tackle/mitigate the most serious concerns
(iv) To improve (i.e., by making more *practical*) Test and Verification plans, and Process Control plans (refer to Process-related FMEA for Construction, Manufacturing, and Subassembly).

Snee and Rodebaugh (2020) also mention three aspects of the FMEA that align the method to Systems Thinking. They pertinently remark as follows: "In 1996, the American Society for Quality defined statistical thinking as a philosophy of learning and action based on the following fundamental principles:

(i) All work occurs in a system of interconnected processes
(ii) Variation [e.g., drop in performance] exists in all processes
(iii) Understanding and reducing variation are keys to success."

They conclude that, "It is those [Systems Thinking] principles that underpin the FMEA approach to analyzing and improving processes". They understood that, "Every system design involves three systems" (Snee and Rodebaugh, 2020), as per Scott's Law number #9 of effective SE (Scott, 2012). Therefore, FMEA probably apply to them all; system being designed, the system used to create it, and where the *deployed* system shall live. A given element (i.e., subsystem) may have an "effect–cause–detection" relationship with other elements within the same system or even across adjacent or "meta" systems.

Carlson further argues the FMEA process is an Engineering Analysis that: (1) is done by a cross-functional, i.e., an eclectic team of subject matter experts; (2) thoroughly analyses product designs and/or manufacturing processes early in the product development process [i.e., design stage]; and (3) Finds and corrects weaknesses before products get into the hands of customers (Carlson, 2012).

Corrective actions arising from FMEA will thus be applied in a descending order of risk. FMEA should not only be applied in the improvement of existing (i.e., previously developed) products and processes. Indeed, following any major failure (e.g., bridge collapse) an FMEA process can be used to perform a "post-mortem" of the event. Furthermore, considering that each infrastructure design, in particular, is expected to surpass (i.e., rectify the inadvertences of) the previous ones, the FMEA process can also be used during the design stage of a new/novel "system" (i.e., large and complex infrastructure) or process (e.g., Construction methodology) to determine failure modes, their effects, and the most efficient approaches to reducing the risk of those failures. It pays reducing risks while still in the design stage.

Carlson (2012) agrees with this position regarding infrastructure design. "An FMEA should be the guide to the development of a complete set of actions that will reduce risk associated with the system, subsystem, and component or manufacturing [or construction] or assembly process to an acceptable level". Such engineering risks will arise from internal damages (to components and/or interfaces) or from adverse external interventions.

> "To analyze the potential design of a new product [or system], design engineers can use FMEA to determine the ways in which the product could [fail or] fall out of specification and become unsuitable for release. They can also analyze the ways in which the product could fail in the hands of the customer [during operations]." (Snee and Rodebaugh, 2020)

Thus, possible failure modes, such as overheated material or improper production line speeds that could drive the product out of specification during production—or trigger dysfunctional, "unsafe" interactions between parts and the environment—are compiled and their causes are determined. Because the product is not in full-scale production, a relevant set of preventive controls and detection mechanisms can be created and audited.

In many organisations, owing to the limited availability of resources needed for these development efforts, a Risk Priority Number (RPN) can be used to determine which areas to tackle and how much effort to invest (i.e., prioritisation of mitigation) in preventing each failure mode and the related out-of-specification or failure condition (Snee and Rodebaugh, 2020). RPN is calculated by working out severity, occurrence, and detection as follows:

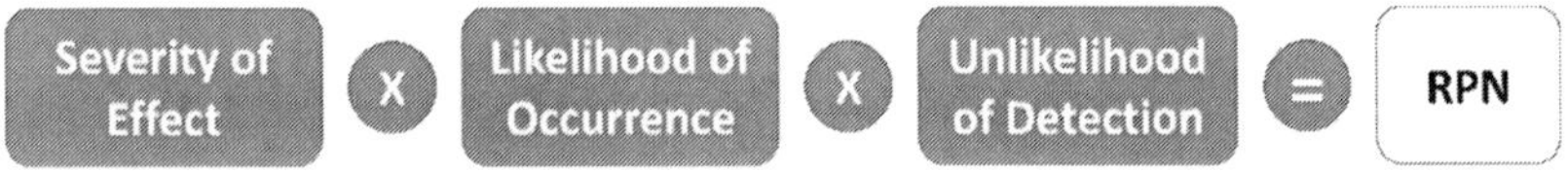

Figure 8.1 – Risk Priority Number (RPN) Calculation Formula

An effect is rated for severity on a scale of 1–10, where 10 is "catastrophic"; for causes of failure, the likelihood of occurrence is rated with a 10 being "highly likely". For detection, the unlikelihood of detecting a cause or failure mode far enough in advance to prevent its effect is rated with a 10 (i.e., nearly impossible to detect before the effect occurs). These numbers are multiplied for each of the several cause–effect–detection combinations for each failure mode to arrive at a *Risk Priority Number* (RPN):

The rating of each of these three elements is performed at the end of the FMEA process, not at the time each element is *first* encountered in the investigation. Thus, the highest effect of a failure mode drives the rating score or RPN. When there is more than one effect for the failure mode, the process of reducing causes of the failure mode will make all effects less frequent (Snee and Rodebaugh, 2020).

In line with Systems Thinking, the definition and behaviours of operational systems, "When analyzing each individual failure mode, it should be remembered that a failure mode may have more than one effect and it may have more than one cause" (Snee and Rodebaugh, 2020)—Multiplicity of causes and effects.

The failure to uncover certain causes can contribute to poor results, since, obviously, no action can be taken against them. Assuming a one-to-one relationship between failures and effects leads to omission of the causes of failure modes—yet, cause identification is at the heart of the FMEA process.

As controls or detection mechanisms are put in place, the RPN should be rescored. When this rescoring takes place, the RPN should be lower as a result of the controls; the lower RPN means that there is less risk to the process. Reduced process risk leads to better (or more stable) performance, which is manifested in various output metrics.

FMEA must not be reduced down to a mere formality; the RPN must prove meaningful:

> "Performing an FMEA just to fill a checkbox in the Product Development Process and then filing it away, never to be seen again, is a waste of time and adds no value. If not for use as guidance through the development process, why waste the time and resources to do it in the first place? If effectively used throughout the product life cycle, it will result in significant improvements to reliability, safety, quality, delivery, and cost." (Carlson, 2012)

Furthermore, because the outcomes of FMEA depend on the FMEA team's subjective knowledge of the process under investigation, the results of the analysis are only as good as that process knowledge. Therefore, the FMEA process should be done by a cross-functional team of subject matter experts (SMEs), Carlson admonishes—and Biswas (2015) provides some practical guidance. Any bias or halo-effect (Stephan, 2013), which occurs if FMEA participants feel obliged to please the facilitator shall be avoided at all costs.

It was noted that the primary objective of an FMEA process is to improve the design, whether to improve an existing product/system or to devise a new one altogether. The FMEA process shall systematically be conducted at three different "levels" as follows:

(1) For **System FMEAs**, the objective is to improve the design of the overall system.

The analysis is done at the highest level of an entire system, which is made up of its various subsystems. When using the Figure 8.3 template, the focus ought to be on *overall* system-related deficiencies, including but not limited to the following:

- System safety and system integration
- Interfaces between subsystems or with other attendant systems
- Interactions between subsystems and/or with the environment
- Single-point failures, i.e., where a single component failure can result in complete failure and/or "total loss" of the entire system (Refer to Module H of the coffee machine system in Section 5.4; a failure of this particular module could have caused total failure)
- Functions and relationships that are exclusive to the system as a whole (i.e., which do not exist at lower levels) and could cause the overall system not to work as intended, e.g., any undesirable behaviours or malfunctions in operations
- Human interactions and service provided.

About human interactions, "Some practitioners separate out human interactions and service into their own respective FMEAs" (Carlson, 2012). Jamshidi (2009) and Gottinger (2012) have further discussed the criticality of *human-interface* in design, which supports this stance.

(2) For **Design FMEAs**, the primary objective is to improve the design of subsystems and/or components—and their "interactions". Accordingly, the analysis is at subsystem or component levels. When completing the template in Figure 8.3, the focus ought to be on product-related deficiencies, with a specific emphasis on the following three aspects:

- Improving the design of the system by enhancing its components
- Ensuring the system's operation is safe and reliable during its useful life
- Improving the design/operation of interfaces between adjacent components. This aspect also concerns human-machine interface

This Design FMEA usually assumes the product (which is the focus of this System FMEA) will be manufactured/assembled according to the right specifications. Nonetheless, systems at times fail partially (or "fail soft") as well as completely, thus entailing an analysis of failure modes for the (complete) system and each of its components.

(3) For **Process FMEAs**, the main objective is to improve the design of the process of construction/manufacturing or assembly (i.e., "How to build"). The analysis is at the manufacturing/assembly process, relying on a "process map" as a step-by-step pictorial sequence of a process, showing process inputs, process outputs, and processing steps, as well as process controls or other enablers. Assuming that the design is satisfactory, the focus would be on manufacturing-related deficiencies. When completing the template in Figure 8.4, the emphasis of FMEA must be on:

- Ensuring the product is built to design requirements (i.e., build-to), in a safe manner, with minimal downtime, scrap and rework
- Improving the manufacturing Construction/Manufacturing and Assembly operations, as well as the transporting of materials
- Integrating the packaging, shipping, incoming parts, storage, conveyors, tool maintenance, storage and labelling, as well as inventory of finished products.

In a nutshell, the Process FMEA seeks to minimise loss of product or performance degradation, as well as assist in developing Preventive Maintenance plans for various "in-service" machinery and equipment. Thus, the higher the complexity of the Solution-System, the more complex and critical Process FMEA will prove.

Every engineer wants to support the accomplishment of safe and reliable systems or products and processes while maintaining happy and loyal customers. As Figure 8.5 suggests, an effective "identification of cause(s) of failure(s)" is at the heart of FMEA processes (i.e., system, design and process levels FMEAs), which should be conducted (repeatedly) in keeping with the system life cycle (ISO 15288) as per Figure 8.2 below. Table 8.1 lists FMEA definitions to consider in completing generic FMEA templates.

This FMEA process is quite powerful. When done correctly, FMEA can anticipate and prevent problems, reduce costs, shorten product development times, and achieve safe and highly reliable products and processes—it helps detect and avert engineering errors.

Nevertheless, RPN is not a perfect representation of the risk associated with a failure mode and associated cause(s). This is owing to RPN being somewhat subjective and statistically "not continuous" (i.e., product of multiplied factors). For the above reasons, it is always recommended that

High Severity items be addressed regardless of their RPN value — This is in line with the Systems Thinking approach to Project Risk Management.

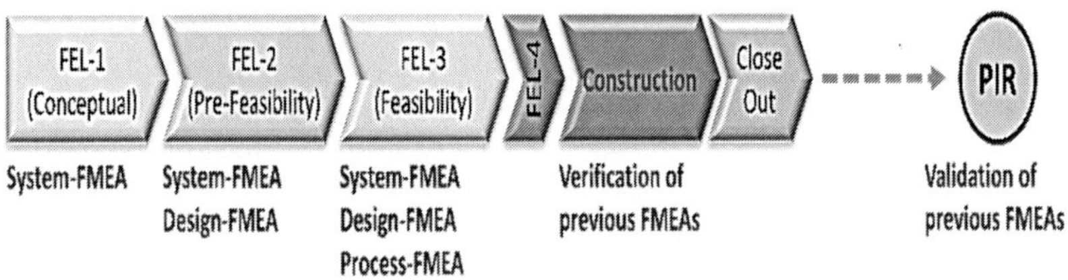

Figure 8.2 – Project Life Cycle Showing Post-Implementation Review

It flows from the foregoing deliberations that FMEA is most comprehensive during FEL-3 (i.e., Feasibility) where both Detailed Design and Execution Planning are complete. Only at that FEL-3 point are FMEA participants in a position to conduct reviews at three levels:

(i) On the whole system under consideration
(ii) On the key and vital "components" of the system
(iii) On the most important process of Manufacturing/Construction of the system

Moreover, some "validation" of FMEA processes ought to take place at the subsequent Post-Implementation Reviews (PIRs). However, conducting these FMEA processes should not preclude the necessity of V&V processes both on the requirements and the system. Jamshidi (2009) also wants such reviews to apply to SoSs (e.g., large and complex infrastructure systems):

> "SoS governance should conduct [FMEA] assessments aimed at evaluating the composite readiness/maturity of the mix of constituent systems and their interaction, as well as the configuration of the SoS in response to the set of defined capabilities and constraints."

For instance, an inquest into the Zeebrugge ferry mishap (of 12 December 1979) revealed: "In this accident, those making decisions about vessel design, harbour design, cargo management, passenger management, traffic scheduling, and vessel operation were unaware of the impact (side effects) of their decisions on the others and the overall impact on the process leading to the ferry accident. Each operated 'reliably' in terms of making decisions based on the information [and approach] they had" (Leveson, 2012).

This *finding* alone attests to the lack of Systems Thinking approach to safety/reliability. "Bottom-up decentralized decision making can lead—and has led—to major accidents in complex sociotechnical systems. Each local decision may be 'correct' in the limited context in which it was made, but lead to an accident when independent decisions and organizational behaviors interact in dysfunctional ways. Safety is a system property, not a component property, and must be controlled at the system level, not the component level [...] Building safer systems requires going beyond the usual focus on component failure and reliability to focus on system hazards and eliminating or reducing their occurrence. This fact has important implications for analysing and designing for [overall system's] safety" (Leveson, 2012)—Thus, FMEA shall follow Systems Thinking.

Accordingly, not only shall FMEA experts and practitioners consider the relationships and/or interconnections between components of a system-of-interest (as well as components of any intersecting systems in their environment or meta-system), but the potential "chains of events" (like *tottering dominos* awaiting to collapse once the first one has fallen) ought to also be considered (i.e., ways of "breaking" those chains) when analysing failure modes.

"When the first domino falls, it automatically knocks down its neighbour and so on until the injury occurs. In any accident sequence, according to this model, ancestry or social environment leads to a fault of a person, which is the proximate reason for an unsafe act or condition (mechanical or physical), which results in an accident, which leads to an injury" (Leveson, 2012)—thus, a minor event might lead to a major accident (owing to non-linearity). In some cases, such chains will entail a series of '*escalating*' and '*balancing*' causal loops; engineering actions (i.e., design) can be used to rectify the scenario by breaking the chain.

The traditional attitude to safety and/or performance failures that focuses on "human errors" is sometimes misleading and unfair. More often than not, operators and users are being blamed for what could easily be identified as "design inadvertences" that could (or rather should) have been detected and rectified during an adequate FMEA exercise during the early phases of the project life cycle—from Conceptual to Pre-Feasibility to Feasibility Phases.

Think about a truck driver who has been driving over a long distance, seated in an *unnatural* position, and forbidden to stop on the road (for a much-needed break) even when his body was impelling him to do so. Should a car crush ensue, must he alone be blamed? It was a Design FMEA failure *first*!

#	Element	Description/Definition
01	Item	An "item" is the focus of the FMEA process. It could be the system itself (in System FMEA), subsystem or component under analysis (in Design FMEA), or one of the specific steps of the manufacturing or assembly process under analysis (in Process FMEA).
02	Function	A "function" is what the item or process is intended to do, usually to a given standard of performance or requirement. It is the primary purpose or design intent of the item (in Design FMEA), or the primary purpose of the manufacturing/assembly operation (in Process FMEA).
03	Failure Mode	The dictionary definition of the word "failure" is: The act of ceasing to function or the state of not functioning [well]. A "failure mode" is the manner in which the item or operation can potentially fail to meet/deliver the intended function and/or its associated requirements. E.g., a high-rise building may collapse (structural failure) or else be rendered "unfit-for-purpose" (functional failure).
04	Effect	An "effect" is the consequence of the failure on the system or end user. This can be a single description of the effect on the top-level system and/or end user, or three levels of effects (local, next higher level, and end effect). For Process FMEAs, consider the effect at the manufacturing or assembly level, as well as at the system or end user.
05	Severity	"Severity" is a ranking number (from 1 to 10) associated with the most serious effect for a given failure mode, based on the criteria from a severity scale, but without regard to the likelihood of occurrence.
06	Cause	A "cause" is the specific reason for the failure, preferably found by asking "why" (5-Whys) until the root cause is determined; beware of "multiplicity of causes and effects". It can arise from a *design deficiency* (Design FMEA) or a *manufacturing/assembly deficiency* (Process FMEA) resulting in failure mode.
07	Occurrence	"Occurrence" is a ranking number (from 1 to 10) associated with the likelihood that the failure mode and its associated cause will be present in the item being analysed. It considers the likelihood of occurrence *during the design life of the product* (in System and Design FMEA), but *during production* (in Process FMEA), without regard to the severity of impact or likelihood of detection.
08	Controls	"Controls" are the methods (i.e., to prevent/detect cause at design) or actions (i.e., at operations) currently planned, or are already in place, to reduce or eliminate the risk associated with each potential cause.

09	Detection	"Detection" is a ranking number (from 1 to 10) associated with the best control from the list of detection-type controls, regardless of severity of impact or likelihood of occurrence.
10	RPN	"RPN" is a numerical ranking of the risk of each potential failure mode/cause; a product of severity, occurrence, and detection (1 < RPN <1000).
11	Actions	"Recommended Actions" are the tasks recommended by the FMEA team to reduce or eliminate the risk associated with potential causes of failure; "Action Taken" is the specific action that is implemented.

Similar ranking numbers shall be used in FMEA (Figure 8.3) and Process (Figure 8.5) Templates

Table 8.1 – FMEA Generic Elements and Definitions

Item	Function	Portential Effect(s) of Failure	Severity	Potential Cause(s) of Failure	Occurrence	Current Design Controls (Prevention)	Current Design Controls (Detection)	Detection	RPN	Recommneded Action(s)	Responsible Person	Actions Taken	Revised Rankings			
											Target Completion Date	Effcetive Completion Date	Severity	Occurrence	Detection	RPN

Legend:

Function	*What the item or process is intended to do, to a given standard of performance or requirement*
Severity	*A ranking number associated with the most serious effect for a given failure mode*
Occurrence	*A ranking number associated with the likelihood that the failure mode and cause will be present*
Detection	*A ranking number associated with the best control from the list of detection-type controls*
RPN	Risk Priority Number — A *numerical ranking of the risk of each potential failure mode/cause It is made up of the arithmetical product of Severity x Occurrence x Detection = PRN*

Figure 8.3 – Failure Modes and Effects Analysis (FMEA) Template (Adapted: Carlson, 2012)

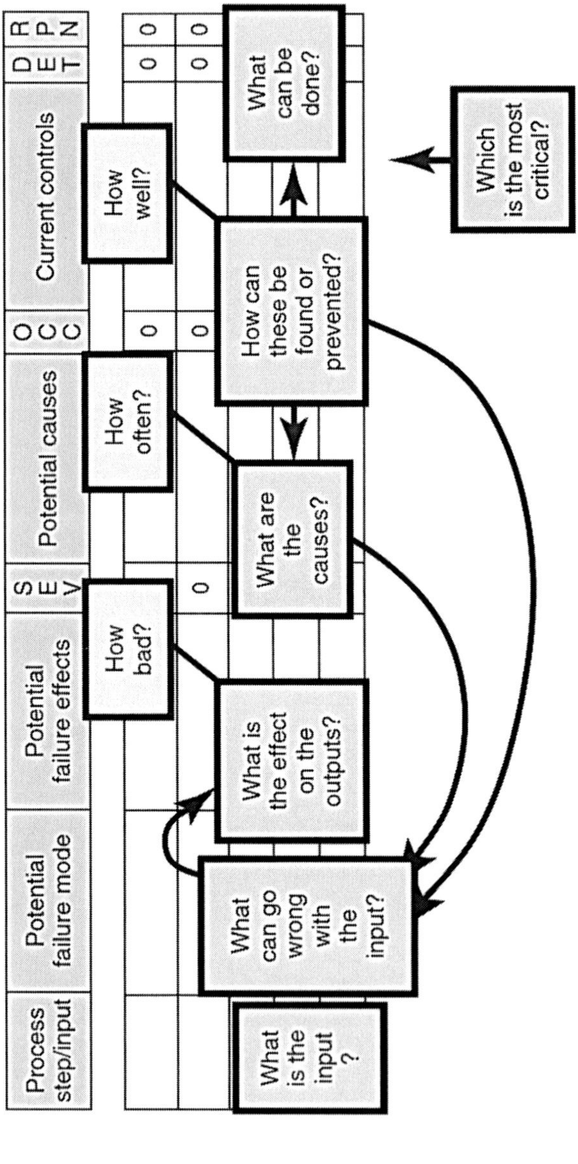

Figure 8.4 – FMEA Generic Process Flow (as per Snee & Rodebaugh, 2020)

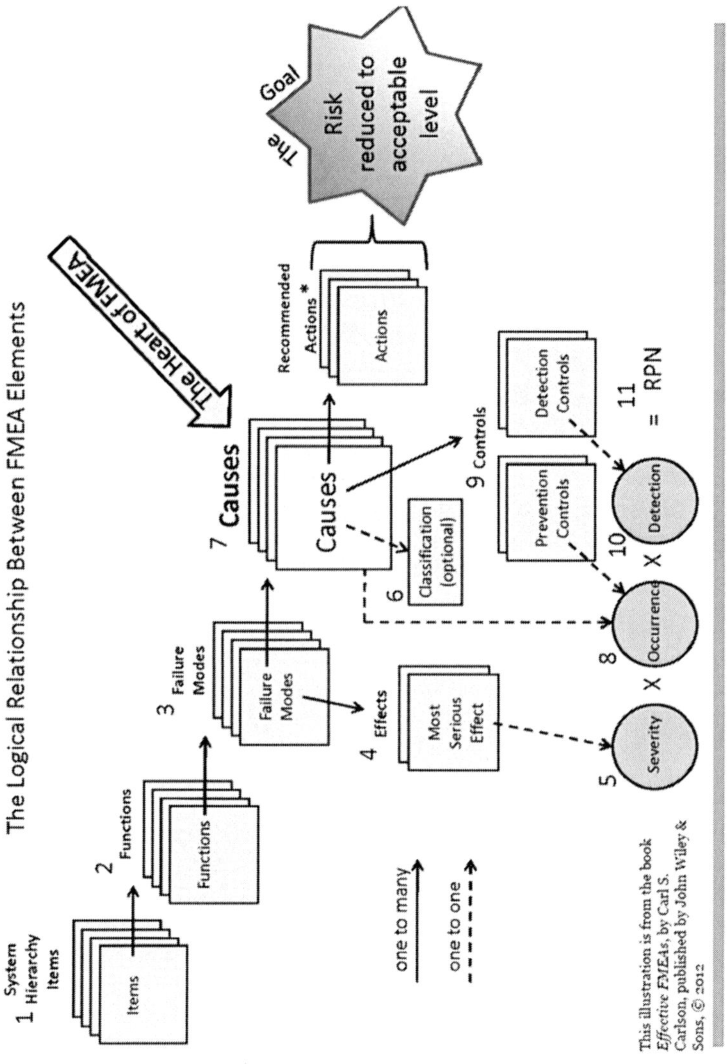

Figure 8.5 – Logical Relationship Between FMEA Elements (as per Carlson, 2012)

The foregoing FMEA process was applied to the engineering design outputs of a Railway Line Electrification project. The outcomes of FMEA processes (i.e., System, Design, and Process) are summarised as follows—for the sake of non-engineer readers, the issues are discussed in an ordinary language:

(1) The rail's System FMEA has focused on the overall/integrated system to reveal that interruptions of electrical power supply and the damage to rails or removal (stealing) thereof, for example, would result in the whole system (i.e., rail operations) failing.

While these failure modes may lead to more or less the same effects, they definitely arise from varied causes (e.g., vandalism or larceny for both, but "burning" damage to equipment for power supply and wreckage for rails). Whereas their severity (i.e., temporary dysfunction/failure of the system) and likelihood proved quite similar, their detection ranking will differ substantially seeing that power interruption is detectable remotely but damages or "removal" of rail lines would require visual inspection.

It was recommended that swingnose crossings (with point motor) be incorporated to mitigate derailments (i.e., larceny triggers) and a computerised monitoring system of the rail "status" be installed, with security patrols and drones deployed to deter theft and vandalism.

(2) The Design FMEA, for example, considered the design aspects of electricity pylons. In customary standards, wooden electricity poles are usually preferred for remote, inaccessible areas due to their reduced costs of installation and/or replacement. The new standard shall upgrade pylon design to galvanised steel irrespective of location to allow for uniformity in their procurement and/or installation.

It is also assumed that effects, severity and detection remained the same as above. However, the high occurrence of cable theft (by way of breaking down the pylons) would turn the proposed design unsuitable (i.e., for repairs), unless such pylons are delivered in prefabricated, easy-to-assemble, and in replaceable units—with enhanced assembly-interfaces. Thus, it was decided that electricity pylon design be changed accordingly.

(3) The Process FMEA similarly considered the construction aspects of the rail line. For instance, the aforesaid high occurrence of rail theft that results in stoppage of traffic (i.e., severity) has incited design engineers to trade the usual "short and bolted" rail segments for "long and

welded" ones that "rail thieves" might find hard to carry off or conceal. while this type is harder to repair in case of derailments, swingnose crossings shall lessen such incidences and the risk of failure will still prove markedly reduced.

These examples exhibit the power of the FMEA process in preventing operability failures. While the FMEA process should apply to most system design, the increasing complexity, deployment, sophistication of the equipment and related processes within safety-critical environments (e.g., nuclear plants, bullet trains, underground buildings, mega-airplanes), makes design errors and system failures more probable, rendering FMEA (more) relevant. The author once had this to say regarding FMEA processes in large and complex infrastructure projects:

> "Failures in these safety-critical environments may have varying effects on the customers, users, operators, communities, as well as on flora and fauna. The effects and severity could range from no harm, including near misses, to devastating harm—be this temporary or permanent, or even loss of life. The strategies available for managing such adverse events usually fall within the purview of traditional risk management that often reduces mitigation to contingency budgets. Yet, little attention is being focused on the available 'proactive methodologies' such as FMEA." (The author)

CHAPTER 9

ENGINEERING MANAGEMENT PLANNING

The process of Engineering Management Planning consolidates most of the information generated thus far to determine the "*Tasks, resources and risks*" involved in the Design and Development of the planned infrastructure system. Design and Development in itself is actually a "phase" of Asset Management. "Asset management as defined in ISO 55000 is a 'coordinated activity of an organization to realize value from assets [i.e., systems]'. This encompasses coordinated and optimized planning, asset [i.e., system] selection, asset acquisition/development [i.e., the EDD], asset utilization, asset care (or maintenance), asset life extension (if applicable) and asset decommissioning/renewal" (Okoh et al, 2016).

> "The physical asset concerned here consists of the systems to be maintained. 'Physical asset itself can be defined as 'equipment, inventory and properties owned by the organization' (ISO 55000)' [...] This would include mechanical systems, electrical and electronics systems, structures and their various components [which make up an infrastructure]." (Okoh et al, 2016)

Maintenance should influence the other phases of the "physical" asset life cycle to contribute to the realisation of a robust physical asset management. "The intention is to use [the findings and insights from] maintenance to promote the integrity of assets across their life cycles in order to uncover more values from the assets, and this is expected to add to the improvement of the asset management (as a whole) of an organization" (Okoh et al, 2016). This makes a lot of sense since, as the author has always advocated, projects are planned for the sake of operations (i.e., "*starting with the end in mind*"). Operational and Maintenance input is needed to inform/guide engineering processes (Mabelo, 2020b).

While Operations and Maintenance teams generally lack project or design experience, "They know very well how a plant should be operating" (Krauss, 2014) and, thus, will be needed to enable operators to take on the variety of tasks they need to *reform* in day-to-day operations; hence, their input is very important to projects. Table 9.1 below specifies maintenance input across the asset life cycle phases; similarly, Figure 9.1 depicts such inputs as regards the Specification and Design stages. The figure clearly indicates that Operations and Maintenance considerations and requirements ought to feed back into the "preceding" Specification and Design stages of the EDD processes. This systemic loop supports the author's dictum of "*Putting Empathy in Operations*" (Mabelo, 2020b).

Design engineers would benefit from involving Operations and Maintenance personnel. "Often the issues arising in a project in the formulation of solutions and the intricate problems [i.e., operability] that require solving in achieving a deliverable [i.e., deployed system] are lost on the operations teams. They have to find their way into the project environment and learn to work with multiple disciplines, engineering [i.e., EDD], procurement and other personnel delivering the outcomes" (Krauss, 2014). For example, the design period for a new high voltage distribution system and transformer was reduced by three weeks (with real cost savings) due to early operations involvement (Krauss, 2014). Therefore, those design engineers involved in Large Infrastructure Projects ought to keep the below Table 9.1 in mind.

Life cycle Phase	Maintenance Influence into Engineering Design and Development
Needs and Feasibility Assessment	Maintenance personnel should always be involved in the systematic process for evaluating and addressing needs, or "gaps" between the current performance of physical assets and desired performance or "wants" [i.e., operational and/or transition requirements] because they usually have significant information about the capacity of existing assets and the comparison of their life cycle costs with those of the desired assets of LIPs.
Engineering and Design	[In the same way as construction personnel are relevant to the "design for constructability" (i.e., the "How to build"),] maintenance personnel are relevant in collaborating with design engineers to promote the "design for maintainability" of physical assets. A lack of maintenance input at this critical phase would prove detrimental.
Procurement Development	Maintenance personnel must collaborate with the procurement [team or] department to ensure quality by verifying and confirming [through V&V] the specifications of spare parts being [procured or] supplied. In addition, maintenance personnel could promote construction and installation for maintainability [or else maintenance may prove costly].

Operation /Utilisation	"Delivering the best value for money in management of physical assets is complex and involves careful consideration and trade-offs between performance, risk and costs over all of the assets' life cycles. There are inherent conflicting drivers to manage, such as short-term versus long-term benefits, expenditures versus performance levels, planned versus unplanned availability" (ISO 55000). Maintenance personnel ought to forge good relations with operations personnel to minimise downtime.
Maintenance	Maintenance personnel possess the technical know-how to retain items or restore them to a working state [i.e., *–ilities*, such as operability status or availability status] in which they shall perform their required functions.
Life Extension (if applicable)	Life extension is sometimes called service or systems life extension, life cycle management or life optimisation [...] the incentives to cut cost and delay the disposal of older systems are attractive [...] risk and cost-benefit analysis must be considered, or else short-term cost savings may be overwhelmed by greater long-term costs, safety problems, or decreased system performance. Maintenance personnel have useful inputs to offer in decision-making process [based on maintain-vs-replace trade-offs].
Decommissioning	As technology advances, assets become obsolete and non-competitive. Any company that does not keep abreast with change may become outperformed by organisations which may develop new monopolies. Maintenance personnel could [or rather should] offer decision support on whether to decommission or not, [and also decide on *disused* items].

Table 9.1 – Maintenance Input into System Life Cycle (Adapted: Okoh et al., 2016)

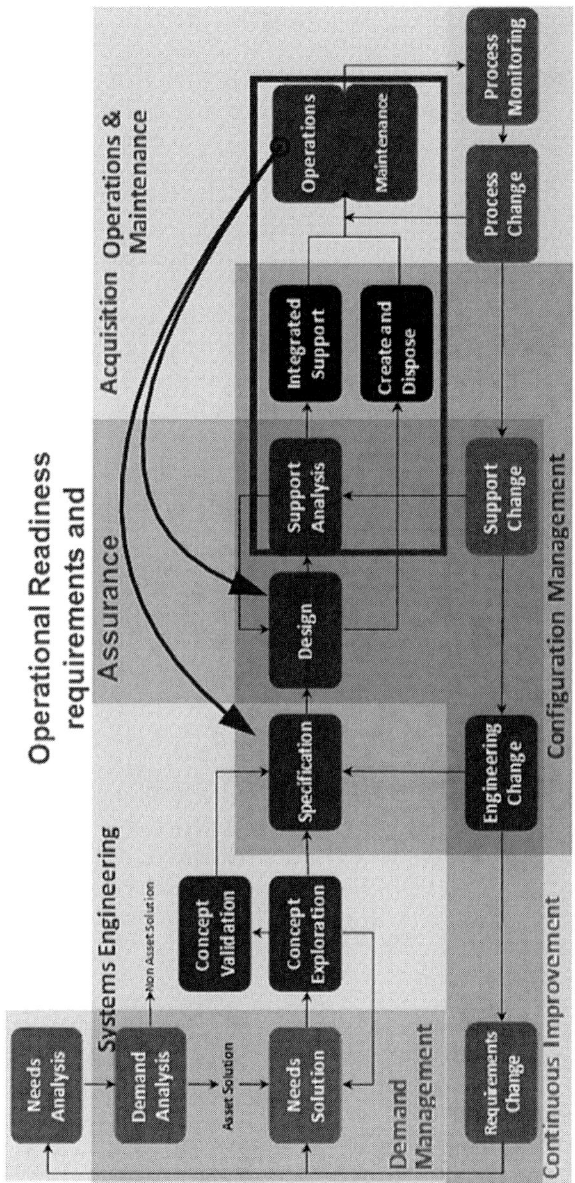

Figure 9.1 – Asset Life Cycle and Operations & Maintenance Inputs (as per Krauss, 2014)

Operations and Maintenance inputs/insights are also crucial in securing the Operational Readiness of the facility (i.e., readiness of the system to safely and effectively transition into operations). In one instance of a plant overhaul, this resulted in an 18% higher project cost, but returned a total saving of 39% in installation, commissioning and start-up costs. Whereas the start-up period was planned to take 8 days, full operation was achieved in 3.5 days, resulting in an additional 4.5 days of production (Krauss, 2014).

The trouble is most Operations and Maintenance personnel are not exposed to project processes, and to LIPs in particular. More often than not, Krauss (2014) bemoans, they would be (and have been) out of their comfort zone and normal sphere of influence and activities. Similarly, project personnel are not always attuned to the operations team and their needs within the project (i.e., they often resent both the people and their works)—both need to learn to work together. Therefore, just as Operations and Maintenance personnel understand their function in the useful lifespan of the system as deployed in its environment, "Engineering Design teams require being competent [not only] in the project delivery, [but also in] the understanding of operation requirements and the connecting of all parts of the project and its deliverables to the business case for the project" (Krauss, 2014).

Furthermore, SE tasks include the inputs and activities of SE personnel in guiding the engineering of the system through all its conceptual and engineering phases. The SE Management process seeks to organise that "technical effort" over the entire project life cycle, whereas the Systems Engineering Management Plan (SEMP) structures the technical endeavour.

In the development of complex systems, e.g., as in LIPs, the most important function of the SEMP is to ensure that all of the many active participants (e.g., subsystem managers, component design engineers, test engineers, business/systems analysts, specialty engineers, subcontractors, etc.) not only know their responsibilities, but also know how they interface with one another—SEMP "gives structure" to the totality of engineering endeavour.

Just as special documentation is required to control system interfaces, so the interfacing of responsibilities and authority within the project must also be defined and controlled. This is generally achieved through preparation and dissemination of the SEMP or its equivalent.

The SEMP is intended to be a living document, starting as an outline, and being further elaborated and otherwise updated as the system development process goes on. Having a formal SEMP also provides a control instrument for comparing [and rectifying] the planned tasks with those accomplished during the project stages (Kossiakoff et al, 2011).

The SEMP documentation, which plans the implementation of SE-related tasks, usually contains a detailed statement of how the SE functions are to be carried out in the course of system development. It can be considered as consisting of four types of planning activities and deliverables as follows:

(1) The **Development Programme/Project Planning and Control** shall describe the SE tasks that ought to be implemented in managing the development programme or (a set of the) *selected* component projects, including the following "document sections":

(A) Statements of Work (SoW)
(B) Allocation of "Roles and Responsibilities"
(C) Reviews, including of scheduling, design, and readiness testing
(D) Technical performance measurements
(E) Monitoring and Control mechanisms

These sections elucidate the "management" of the entire engineering endeavour. It has transpired from Figure 6.4 (back in Chapter 06) that the "Scope" of the engineering/technical efforts (and its supporting activities) shall comprise *approved* "work packages", e.g., (1) Project Management, (2) Monitoring and Control, (3) Technical Management, (4) Environmental Management, (5) Procurement Management, (6) Project Support, and any others as applicable.

From these work packages, the project manager will be able to develop the "Plan" for the implementation of the project under consideration. Further, any Monitoring and Control regime has always to rely on a *baselined* plan (i.e., approved and frozen). Such a "Baseline Plan" should, thus, include the following project delivery aspects:

(i) Listing of activities and/or deliverables
(ii) Organisational structure and resources (i.e., skills) required
(iii) Schedule and/or timelines (e.g., key milestones)
(iv) Estimated budget (i.e., costs) and cash flow (where applicable)
(v) Major risks involved and proposed mitigation steps.

In addressing the above implementation aspects, the engineering manager would definitively depart from traditional approaches and might be able to "*establish or adhere to a realistic design programme*", and ultimately prevent the unfortunate situation of "commencing construction *before design is complete*" which is always a recipe for disaster, as most Large Infrastructure Projects (LIPs) experts maintain.

(2) The **Systems Engineering Process** describes the SE process as it applies to the development of the infrastructure system, such (a) Operational Requirements, (b) Functional Analysis, (c) Systems Analysis and Trade-off Strategy, and (d) System Test and Evaluation Strategy, including any possible need for a mathematical and/or computer modelling and/or prototyping—whether it be large or small in scale.

Thus, the FMEAs will ensure the right system is delivered that provides the solution which shall satisfy the criteria: "*Meet requirements to a maximum possible extent*".

(3) The **Engineering Specialty Integration** seeks to describe how areas of specialty engineering should be integrated into the primary system design and development. These include (a) Reliability, Maintainability, Availability, Operability, Durability Engineering—*RAMOS*, or *—ilities* Engineering that addresses the *qualitative* fitness-for-use criteria, (b) Constructability/Producibility Engineering, (c) Safety Engineering and (d) Human Factors Engineering.

This section focuses on system aspects relating to *—ilities* Engineering. These mostly "emergent" characteristics pertain to the "integration" of the system construed as a CAS/CASoS during Design, Production, and Operation. An integration failure in any such stages might render the system futile, if not deleterious (i.e., safety failure) in future operations.

(4) The **Engineering Development Guidelines** seek to describe how critical outputs are progressively elaborated throughout the life cycle and recommend suitable levels of definition in line with project phases. As a rule, the indicative accuracy of the project estimates (e.g., budgets of cost and time) increases with the level of engineering elaboration. Table 9.2 provides a "generic" Engineering Development Table that project leadership (with the engineering teams providing a substantial input) might adapt to suit the specific circumstances of their project(s).

Kossiakoff recommends that, "The steps in the development of a new system can be thought of as an orderly progressive 'materialization' of the system from an abstract need to an [adequate] assemblage of actual components cooperating to perform a set of complex functions to fulfil that need" (Kossiakoff et al, 2011).

While these four sections are important, only the first three are specific to the project at hand. The last section generally serves as a "yardstick" to assess the effectiveness of the main ones. Nevertheless, it is Section 4: Guidelines (as approved by project leadership) that proves most challenging to engineering designers, if not to the entire project team.

Considering that Table 9.2 merely provides guidelines for application across the organisation as part of their life cycle methodology, the onus is on the project team, if not on the project management office (PMO) to take a clue from the guidelines and propose *target* values or parameters (to be approved by project leadership), given the specific circumstances of the project (e.g., scale and complexity level, impact and/or criticality to operations).

A further, equally thorny challenge arises from the ability of the engineering team to provide assurance and verification that they have actually met the required level of definition; for example, "*70% Engineering at Feasibility*". Alternatively, "*What resources will be required to reach 70% engineering?*" In terms of planning, such crucial questions ought to largely be answered by effectively implementing the other SEMP-related sections, namely Sections (1), (2) and (3).

Further, Ferguson cautions: "Think carefully about the resources available and needed—the phrase capacity [i.e., enough resources] and capability [i.e., right skills] is useful in this [EDD] context" (Ferguson, 2014). Engineering Planning shall address the availability of resources to produce the required artefacts (e.g., drawings, plans, reports).

Whereas developing a "Project Planning and Control" regimen is important in managing the engineering endeavour, it is worth noting that the "Systems Engineering Process", which produces the envisaged "solution", is actually at the centre of the SEMP process. There is no glory in suitably managing the engineering process *per se*, only to fail in developing the envisaged system. Such a scenario will probably resemble the proverbial, "*It was a great operation [i.e., followed process], but unfortunately the patient died*"!

A team of young engineers nearly fell into that pit. They were appointed to design a plant that will use natural gas from Angola to produce 2000 metric tons of ammonia per day (at 98% purity, by weight). They proceeded to conduct a literature review to grasp what industry states as best practices —this entailed extensive research of recommended production conditions.

Engineering Definition "Guidelines" and Cost Estimates pertaining to FEL Phases					
Project Lifecycle Classification	**FEL-1**		**FEL-2**	**FEL-3**	**FEL-4**
Usual Description	Conceptual Study		Pre-Feasibility Study	Feasibility Study	Pre-Construction Phase
Class of Estimates	Level 1		Level 2	Level 3	Level 4
Gate Review Level	Gate 1		Gate 2	Gate 3	Gate 4
End Use (Typical purpose of Cost Estimates)	Concept Screening; comparison of alternatives; Preliminary investigation.	Concept Study; Identification of requirements and cost of level 2 study.	Budgetary, Viability; Determine validity of business case.	Appropriation or Control; Detailed control baseline for monitoring variation to budget	Check Estimate; detailed validation of discrete parts of total budget.
Level of Engineering Definition	**0% to 1%**	**1% to 5%**	**10% to 30%**	**40% to 70%** [Prototype, if any]	**60% to 100%**
Level of Contingency	30% or more	30% to 50%	20% to 30%	10% to 15% based on Q_nRA	Up to 10% based on Q_nRA
Indicative Probability Range	No better than 80%		Not greater than 90%	Not greater than 90%	Not greater than 90%
Indicative Accuracy Range	-50% to +50%		-20% to +20%	-10 to + 15%	-5% to + 10%

Table 9.2 – Guidelines for Engineering Design and Development

Being armed with the required (theoretical and practical) background, they concluded that the design was to be broken up into different segments; namely, (1) Steam Reforming, (2) CO_2 Removal, and (3) Methanation. While Steam Reforming mostly takes place through the use of a primary and secondary "reformer", with the aim of converting the natural gas into a hydrogen-rich product, the CO_2 Removal was required to remove any carbon dioxide (CO_2) that *inopportunely* acts as a catalyst-deactivator in the Methanation step. Methanation entails a refrigeration cycle to create the final ammonia product.

Having designed the entire process, the team used a "simulator software" to simulate the production of ammonia under industry conditions. However, the simulation failed in accurately "sizing" a reactor to be used in the CO_2 removal process. As expected, blind reliance on computerised simulations (*with no sense of reality*) led them to design errors.

This design error led to an impossible ratio of CO_2 entering the Methanation step and, as a result, causing impossibly low (- 400°C) cooling temperatures. Further, this caused an *inaccuracy* in the composition of the final product. The "reactor" was also unfeasible as it was sized at a height of 50 m and a diameter of 14 m; these dimensions were massively out of range.

To correct this design error, the team had to go back to the first principles of Detailed Engineering, which meant redesigning the "reactor" using well-understood tools and formulas, "*where one could feel each step of the process*". They found out which of their initial assumptions were flawed and/or impractical—This came from understanding one's design.

> "Our designs [as rectified based on understanding] were more intellectually manageable, and the potential interactions among components could be thoroughly planned, understood, anticipated, and guarded against […] testing was possible and could be used to eliminate [design] errors before use." (Leveson, 2012)

This approach resulted in better, more realistic values. The "reactor" was then sized at a height of 7.00 m with a 2.20 m average diameter, which was more aligned to both design specifications and industry best practices. The main learning, again, was that technology or processes are only good when there is *understanding* of what happens at the back end. Practical knowledge of first principles is crucial to any Detailed Design. These young chemical engineers learnt it the hard way—what a way to earn your stripe of wisdom?

This near-blunder could have compromised the whole ammonia project. No wonder Al-Mazrooie cautions, "In complex projects where activities are closely related to each other, the longer it takes to find the mistake, the more additional work can be created in the process of correcting the mistake and the more the total project performance can be degraded" (Mazrooie and Bajracharya, 2013).

In a different scenario that will probably result in a happier ending, a team of engineers enlisted assistance from the author to guide them through their engineering endeavour. The project consisted of designing a Sport Hub at a local university. The planned facility encompassed Olympic track fields and a swimming pool, in addition to a soccer/rugby stadium, a gymnasium, basketball courts, tennis courts, yoga studios, cricket fields and a clubhouse. The complete "Sport Precinct" was to include the following amenities:

(i) Business/Primary Requirements:

"*A sport precinct where students could congregate for various sporting activities*".

(ii) Proposed Concept of Operations:

"*Up to 2,000 students should be able to walk to the precinct situated at least half a mile away from auditoriums, laboratories and libraries where they should practise and celebrate sporting activities without disturbing teaching and learning sessions on the campus, no matter at what time of the day or the weather conditions*".

(iii) Additional Stakeholders Requirements:

- Safety Requirements – *i.e., shelter against the elements, protection against drowning, unrestricted access to health & safety personnel (e.g., ambulance, fire-truck).*
- Environmental – *i.e., minimise usage (i.e., waste) of potable water, reduced/controlled use of hard chemicals (e.g., fertilisers, pool cleaners), noise barriers, waste removal.*
- Transportation – *i.e., adequate parking bays, safe connection to the main road.*

(iv) Authorised/Approved System Requirements:

"*University Board has voted that the proposed requirements be fully satisfy*".

The engineering team further committed to seeking innovation in design. Thus, they *first of all* considered that, "Good design practice will always require a combination of basic engineering principles, experience and judgment in order to furnish the best possible structure [of a system] to suit an individual site within reasonable economic limitations; it does not preclude justifiable exceptions [...] provided the exceptions [i.e., innovative designs] are based on sound engineering principles" (NJDOT, 2016).

Having considered these project objectives and scope, the engineering team proceeded to develop a "breakdown" of authorised/approved requirements, the RBS (see Table 9.3). Owing to space limitations, only the "swimming pool" portion of the Sport Hub project scope is reflected in the subsequent sections to illustrate the Engineering Planning process.

From this hierarchy of requirements (RBS), they conducted a Requirements Analysis to categorise the requirements into measurable System Requirements and the required functions for the swimming pool, considering the *ring-fenced* scope. These functions were then applied to a Functional Allocation process as per the template provided in Figure 2.5 above to produce a "combination of functions" to be allocated to subsystems as "function enablers". This will determine the "architecture" as illustrated in Figure 9.2.

In practical life, such functions should have reflected not only the activities necessary to satisfy defined operational objectives in the intended environments, but also activities pertaining to the support, maintenance, logistics, and disposal of this swimming pool.

At this point in the EDD process, the team also thought it relevant to apply the Hamburger Model (i.e., Problems/Requirements versus Solutions), as well as the Context Diagram to define the required devices (i.e., subsystems and/or processes) and their interfaces with one another and with their surrounding environments. This is because these techniques are intended to supplement, not oppose each other—although not *graphically* shown here.

For example, when considering the requirement for a cost-effective heating of water, the Hamburger Model provided the team with trade-offs between two technical solutions, i.e., a standalone electric heater and a solar-heated pool. While both "devices" might satisfy that specific requirement, the demand for ecological protection (i.e., renewable energy) prevailed to tilt the decision towards choosing a "roof-mounted solar heater".

Req 1.1 Shall provide Olympic Swimming Pool	**Req 1.1.1** Shall facilitate swimming	• **[Req 1.1.1.1]** Shall fill pool with water at rate of "x" litres per sec • **[Req 1.1.1.2]** Shall maintain pool "swimlanes" within ± "y" cm of initial position • **[Req 1.1.1.3]** Shall keep pool water at no current or draft
	Req 1.1.2 Shall keep pool water clean & fresh	• **[Req 1.1.2.1]** Shall filter pool water at rate of "abc" litres per sec • **[Req 1.1.2.2]** Shall recycle pool water at rate of "xyz" litres per sec • **[Req 1.1.2.3]** Shall prevent dirt & impurities (> "z" mm) from entering pool
	Req 1.1.3 Shall provide safety protection	• **[Req 1.1.3.1]** Shall prevent any individual from falling into pool (when not in use) • **[Req 1.1.3.2]** Shall provide SANS approved (Std "123pq") rescue fixtures & Features • **[Req 1.1.3.3]** Shall allow for drainage/emptying of pool water at rate of "xyz" litres per sec
	Req 1.1.8 Shall provide Pool Housing	• **[Req 1.1.8.1]** Shall provide solid housing (i.e., enclosure) to pool as per SANS ("Std 456pq") • **[Req 1.1.8.2]** Shall provide roof to pool housing as per SANS ("Std 789pq") • **[Req 1.1.8.3]** Shall remove installed roof (as per **[Req 1.1.8.2]**) in hot days (if temperature >22°) • **[Req 1.1.8.4]** Shall provide Lighting in pool enclosure as per SANS ("Std xyz123") • **[Req 1.1.8.4]** Shall provide ventilation in pool enclosure as per SANS ("Std xyz456")
	Req 1.1.9 Shall provide Water Heating	• **[Req 1.1.9.1]** Shall heat up water to ± "tt" degree Celsius (° C) in cold days (if temperature <15°) • **[Req 1.1.9.2]** Shall afford cost-effective (i.e., $ "z" per ° C per sec) heating of pool water

Requirements are kept "high-level"
for simplicity of illustration purposes

Table 9.3 – Example of Requirements Breakdown Structure (RBS) for the Swimming Pool

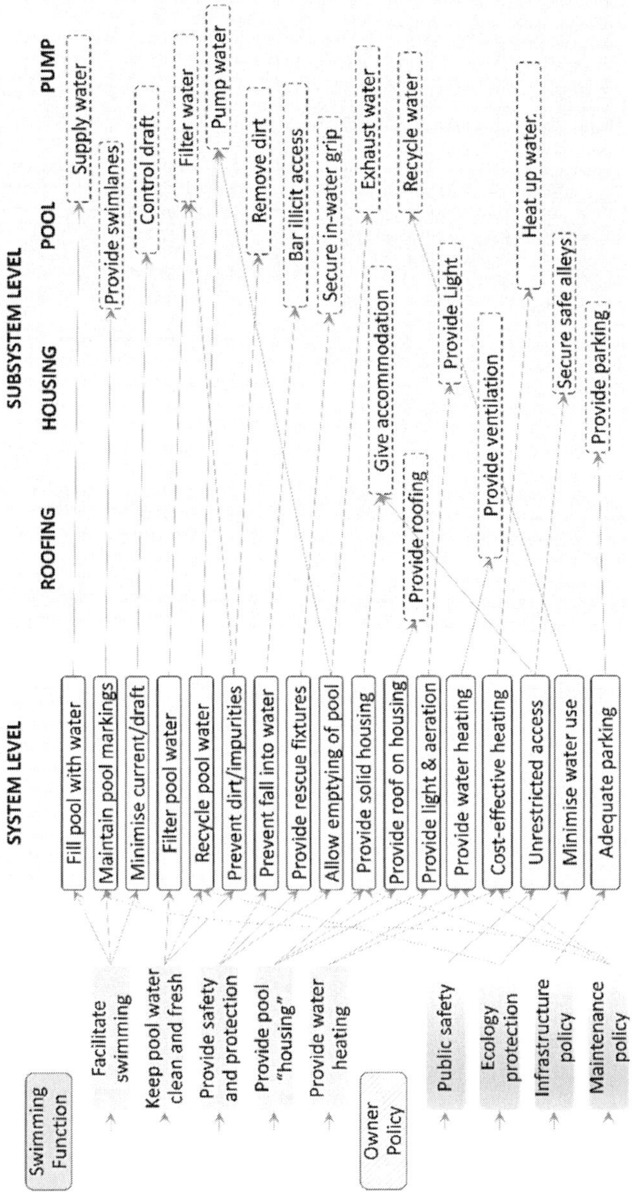

Figure 9.2 – Example of Functional Analysis and Allocation for Swimming Pool System

Likewise, when considering the pool and its surrounding space vis-à-vis the need to prevent spectators from "*falling into the pool and drowning*", the team looked at the Context Diagram to safely manage that key "interface" by installing a steel balustrading around the pool which proved practicable, cost-effective, as well as easy on maintenance.

Due to space constraints, both the Hamburger Model and the Context Diagram for this "swimming pool" system are not included here. These considerations, among many others also not illustrated here, helped the team to develop the subsequent "breakdowns" of facility and system, namely the FBS and SBS (see Figure 9.3 and Figure 9.4 below).

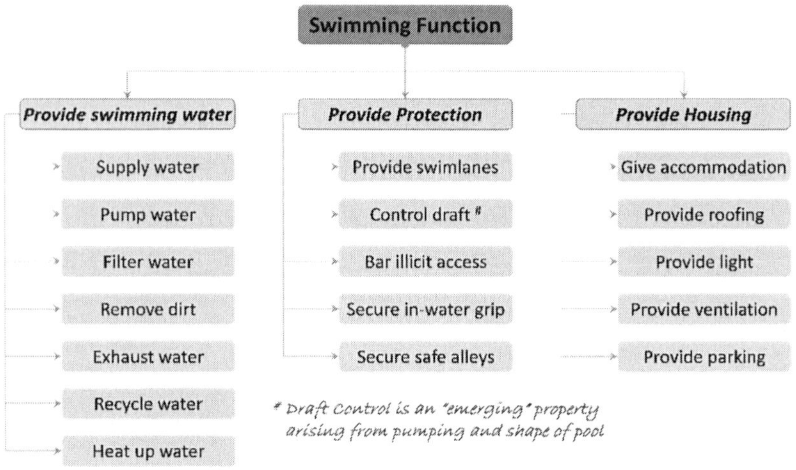

Figure 9.3 – Example of FBS for Swimming Pool System

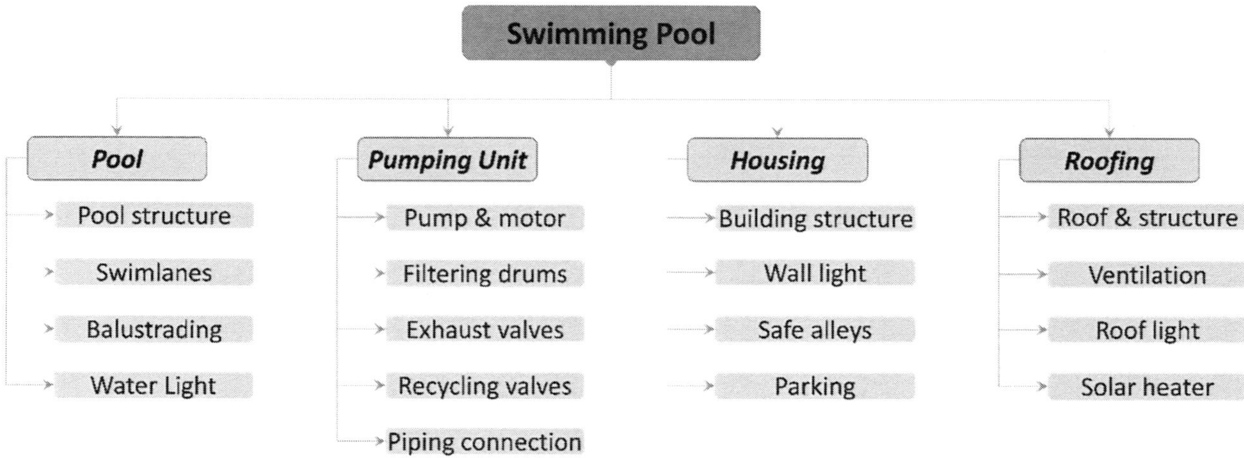

Figure 9.4 – Example of SBS for Swimming Pool System

House of Quality

Quality Function Deployment for "Swimming Pool Design"

■	Strong & Positive
▲	Weak & Positive
▲	Weak & Negative
■	Strong & Negative
—	No Connection

		Importance (Scale of 1 to 5)	Relative Importance (in %)	Relative Ranking	Appropriate Plastering/Tiling ↑	Flat Pool Bed ↑	Rectangular Shape ↑	Spread of Inlet/Outlet ↑	Sunlight Access ↑	Filtering Power ↑	Acceptable Impurity Level ↓	Water Bleaching ↓	Benchmarking: Old Campus Swimming Pool	Benchmarking: Private Swimming Pool
01	Smoothness of Surfaces	5	20.8%	01	9	3	2			6	7	6	2	4
02	No Sharp Angle	1	04.2%	07	8	7	7	3		2	6		2	4
03	Steady Depth	4	16.7%	03	3	9	6		3	6	3	1	4	5
04	Adequately Wide	4	16.7%	03	2	2	8	6	6	7	6	6	3	4
05	Concealed Water Inlet/Outlet	2	08.3%	06	8		7	9	1	7	5	7	4	3
06	Natural Lighting	3	12.5%	05	6	4	5		9	7	7	4	5	4
07	Clean and Fresh Water	5	20.8%	01	8	6	2	8	4	9	8	8	4	5
	Technical Difficulty Involved				2	3	2	3	5	3	2	2		
	Measurements Unit				mm	degree	L/w	no/m	m^2	KW/m^3	Kg/m^3	Kg/m^3		
	Object Target Value				5	0	2	0.04	500	0.5	0.2	0.01		
	Absolute Importance of Design Parameter				06.1	04.5	04.7	03.5	03.5	06.8	06.2	05.2	24	29
	Relative Importance of Design Parameter				3	6	5	7	7	1	2	4		

Figure 9.5 – Example of HoQ for Swimming Pool System

Further, the engineering team proceeded to the design phase and, while still following the Dutch MIE Model, elected to also incorporate the QFD with its four Houses of Quality or HoQs (i.e., only for the "Pool Structure"). This tool assisted in benchmarking customer requirements against competition and, even more significantly, in defining the relevant technical characteristics of the facility, its parts, and construction processes and activities.

It transpires from the "HoQ for Planning" (of the Swimming Pool) that the planned pool might score a higher ranking than most competition regarding "Smoothness of surface", "Adequately wide", "Clean and fresh water". This HoQ also indicated priorities with regards to Engineering Characteristics: (1) Filtering power, (2) Acceptable level of impurities, (3) Appropriate plastering, and (4) Water bleaching. Accordingly, the engineering team went on to develop the remainder of Houses of Quality, the other HoQs.

Altogether, these steps provided the team with sufficient information for the development of the "breakdown" of works (WBS) for project implementation. It is worth noting that these "works" are not complete without the pertinent findings and insights from the Maintenance, Environmental, Procurement, and Support divisions. Furthermore, the "works" ought to align with the applicable project life cycle methodology by indicating *what* "activities" should take place *by when* and *by whom* they are to be done.

Figure 9.6 below seeks to summarise the WBS for this swimming pool portion of the project. This WBS proved instrumental in determining the "resources" (i.e., personnel, timelines, and cost budgets) needed for the successful implementation of the project. In so doing, the engineering team was empowered to "*establish or adhere to a realistic design programme*", which most traditional engineering usually fails to attain. In addition, when translated into a Gantt Chart (time-lined representation of activities) and S-Curve (cash-flowed representation of costs), the *programme* should make it feasible to monitor and control the project works.

Unfortunately, there might still be those instances of project teams failing to manage the engineering endeavour and, thus, will end up squandering precious man-hours. This occurs when the team overspend on resources or over-elaborate on design. Completing the "design" prematurely (i.e., too early) in the project life cycle is a typical example of such.

The author was once taken aback during the FEL-2 (Pre-Feasibility) gateway review of a port project when the engineering team boasted of having completed 95% of the design for all the four options, and had produced drawings that were quasi-ready for construction. The author's shock was justified since most life cycle methodologies for megaprojects recommend that FEL-2 should address Functional Allocation and Subsystem Definition, and finalise "*optioneering*" (i.e., matching of options against each other to select the most viable one). FEL-2 usually entails 10% to 30% of the engineering efforts being expended.

Since the Measurements Domain (i.e., shape, size and technical characters) was affected at "subsystem" *instead of* at "element" level, the accuracy of the estimates (e.g., costs of project implementation) was likely to be around -20% to +20% at FEL-2 phase. Further, by prematurely executing 95% engineering at that stage (and they did so "*for all four identified options*"). This was bound to lead to two engineering inadvertences as follows:

I. Finalising Engineering Design at FEL-2, before the Architecture-domain is fully explored (i.e., assemblies and elements finalised) creates the risk of a flawed design (due to flawed or incomplete architecture) being rendered into "Issued for Construction" drawings.
II. At least three out of these four proposed options will in due course be discarded, seeing that only the "*single-go-forward*" option (i.e., one out of four options) would ever be taken to the ensuing FEL-3 (Feasibility) phase. (This will simply mean that the engineering effort expended on those three discarded options have been wasted.)

The above issues of mismanagement of engineering efforts are not the only challenges that engineering teams are faced with during the course of project delivery. Indeed, in light of complexity dynamics that play themselves out in LIPs, the "works" (or at least at the level of "work packages") should be treated as *interconnected* elements that interact with each other and with the environment outside of the project team. Treating project "work packages" as unrelated items will deprive the engineering team of the opportunity of using "high-leverage" items (when identified) to engender positive *ripple effects* across the Realisation-System, should they be put to effect.

This is about complexity:

> "Construction projects [e.g., LIPs] are faced with a challenge that must not be underestimated. These [large infrastructure] projects are increasingly becoming highly competitive, more complex, and difficult to manage. They become problems that are difficult to solve using traditional approaches". (Maqsood et al, 2009)

Cristobal takes a similar view as follows; "Structural complexity stems from large-scale projects [i.e., LIPs] which are typically broken down into small tasks [i.e., work packages] and separate contracts. Projects [...] where the complexity stems from the difficulty in managing and keeping track of the high number of interconnected tasks and activities [i.e., networks of works] are likely to have this type of complexity" (San Cristobal et al, 2018). Here could be a good place for the engineering team to recall Rechtin's advice:

> "Choosing the appropriate aggregation of [selected] functions [as part of the Functional Allocation] is critical in the design of systems [as CASoS] […] in partitioning, choose the elements so that they are as independent as possible, that is, elements with low external complexity and high internal complexity [e.g., far higher complexity in engine than in associated drive shaft] […] choose a configuration in which local activity is 'high speed' and global activity is 'slow change' […] [and entails] minimal communications between subsystems." (Maier and Rechtin, 2009)

Yet another practical guidance comes from none other than Ryschkewitsch:

> "Neither the world's greatest design, poorly implemented—nor a poor design, brilliantly implemented—is worth having." (Ryschkewitsch et al, 2009)

Since both delivery aspects are necessary, Construction (i.e., "How to build") shall be carefully planned, just as Design (i.e., "What to build").

> "Systems engineering is holistic and integrative. It incorporates and balances the [main] contributions of structural, mechanical, electrical, software, systems safety, and power engineers, plus many others, to produce a coherent whole […] Systems engineering is not only about the details of requirements and interfaces among subsystems […] Systems engineering is first and foremost about getting the right design [i.e., "What to build"] – and then about maintaining and enhancing its technical integrity, as well as managing complexity with good processes [i.e., How to build", in terms of 1y/Integration activities] to get it right." (Ryschkewitsch et al, 2009)

A Project Execution Plan (PEP) shall assist in integrating those disciplines:

> "[The SEMP is] [a] set of activities which control the overall design, development, [while the PEP controls] implementation and integration of a complex set of interacting components or systems to meet the needs of all the users." (Kasser, 2015b; citing DERA, 1997)

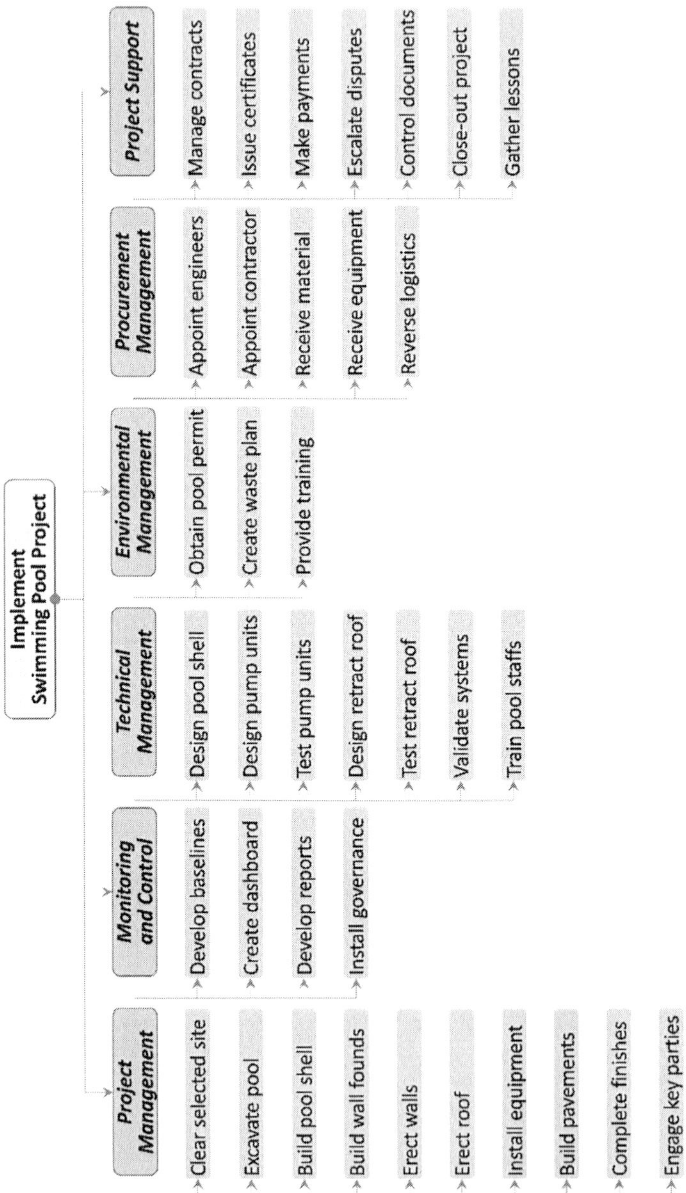

Figure 9.6 – Example of WBS for Swimming Pool System

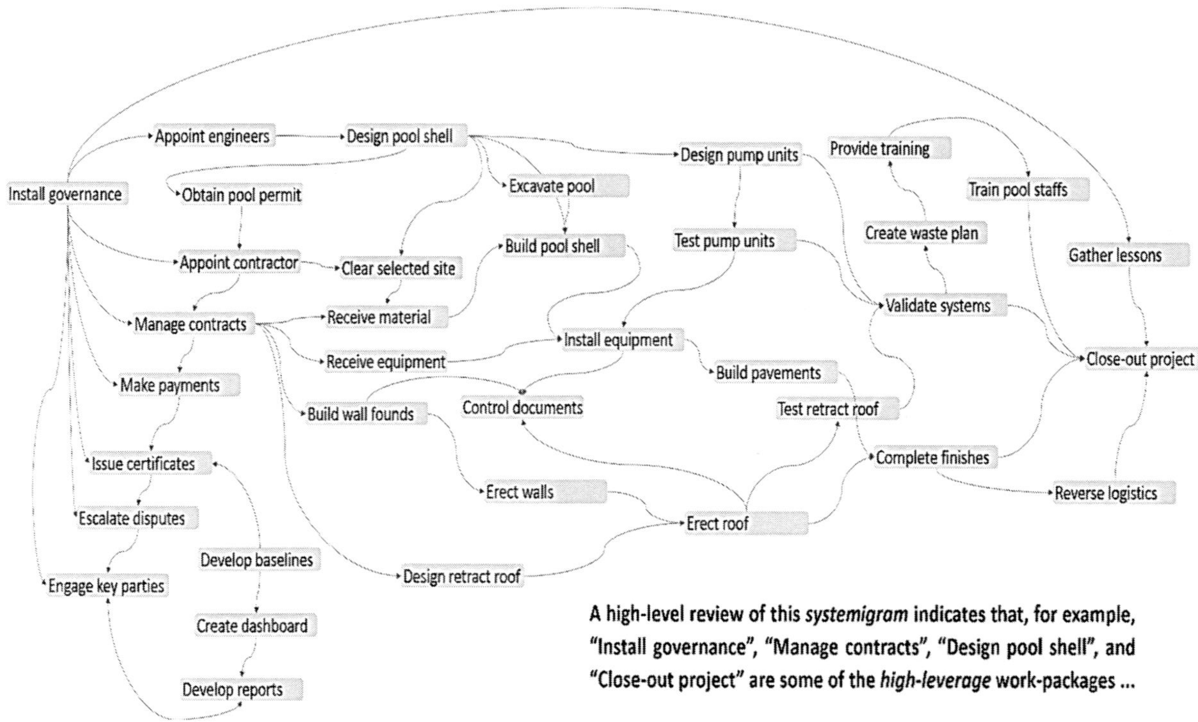

Figure 9.7 – Example of Systemigram for Swimming Pool System

While the complexity of the Swimming Pool might not justify applying such a tool, the "*network of work packages*" (or systemigram, an outline showing how a system works, or *relationships* between the parts of a whole) is included in Figure 9.7 for illustration purposes. However, since this simple example is based on limited data, readers are not encouraged to further analyse this illustrative systemigram or to infer any generic interpretations, but instead explore this useful tool for application in their megaprojects. This is because:

> "Well-structured complex problems consist of a set of interconnected well-structured non-complex problems and since the remedy to one ["work package"] may affect another, these problems [i.e., design of infrastructure systems] cannot be solved in one pass through the problem solving process [i.e., the Engineering Design and Development process]." (Kasser, 2015b)

> "[A]n adaptive systems approach to decision-making has to be closely coupled with the integrative framework to embrace complexity and high uncertainty. The approach is adaptive because it acknowledges the dynamic nature of systems, requires continues monitoring, and flourishes on active learning with an understanding of human decision-making processes." (Jamshidi, 2009)

An understanding of a systemigram assists in decision-making, e.g., by inferring how any change to a particular "work package" might affect the remainder of the project. In the example of the Swimming Pool sub-project, it would appear that the following "work packages" have high-leverage vis-á-vis Engineering Design and Development:

(i) "Install (Project) Governance"
(ii) "Manage (Project) Contracts"
(iii) "Design Pool Shell"
(iv) "Close-Out Project"

The above being the case, what would a failure of the "Manage (Project) Contracts" work-package produce to this Swimming Pool sub-project? Just imagine the effects of having some incompetent project manager assigned to that particular work-package. Similarly, what will the implications be of *mismanaging* the "Design Pool Shell" work-package? "In megaprojects, wholesale changes in design during execution are usually impossible. If the team does not act on the information, no cost growth is generated, but operability [of the system] is affected, often in devastating ways" (Merrow, 2011). The identification and effective management of high-leverage "work packages" must not be underestimated—or else the intended "*improvements of operational environment*" will suffer.

CHAPTER 10

ENGINEERING DESIGN AND OTHER PM PROCESSES

The Project Management Body of Knowledge (known as PMBoK) states; "Projects create value in the form of improved business processes [i.e., operations], are indispensable in the development of new products [i.e., systems] and services, and make it easier for companies to respond to changes in the environment, competition, and the marketplace. Successful delivery of large and complex infrastructure shall conform to effective project management […] processes" (Project Management Institute, 2013).

The PMBoK further argues that:

"Project management processes and product-oriented processes [that create the product or service] overlap and interact throughout the life of a project" (Project Management Institute, 2013). Since EDD concerns itself with devising a new product or system, it should be expected that engineering processes will overlap and interact with PM processes when it comes to large and complex projects. Moreover, there has been a rising appreciation and attention to the notion of project complexity (San Cristobal et al, 2018).

Therefore, the subsequent sections will discuss to which extent critical PM processes overlap and interact with EDD. Such interfaces include processes relating respectively to Prototyping, Procurement, and Monitoring and Control. Those discussions (on interfaces) will revolve around the rationale and effective methods of applying such processes to infrastructure projects.

10.1 Engineering Development and Prototyping

The basic-to-detail Design Process takes place during Feasibility (FEL-3) and should go through stages: (1) derivation of the Unit's requirements (at Development Specification), (2) Requirements Verification, (3) proposing of Workable Unit and Assembly Concepts, and (4) Testing of Prototypes.

In the development of a new complex system, decisions regarding which subsystems and components might require further development and testing prior to their "full-scale" engineering, and issues regarding their physical implementation, are frequently more difficult and critical (to operability of LIPs) in view of any undesirable "emergent behaviours" than those related to their functional design and/or performance (Kossiakoff et al, 2011).

The envisaged testing will often entail the prototyping of the said system. Kossiakoff argues that, "System elements involving dynamic behavior with feedback may be analyzed through simulation [e.g., using computer models] but usually require the construction and testing of experimental models to establish a firm basis for engineering" (Kossiakoff et al, 2011).

When the physical characteristics (e.g., fatigue cracking, deflections under wind loading) would not lend themselves to analysis or simulation, prototyping will require the "critical element" to be designed and built (in small or large scale) and tested to reveal potential "emergence" problems. Hence, the designer asks:

(1) What elements/aspects could go wrong—what will their impact be?
(2) How will they first manifest themselves, and what are their triggers?
(3) What could then be done to address and make those things right?

Prototyping thus affords a proxy of the validation of the "whole subsystem", which assists in the mitigating of the risk of failure—whether total or partial.

User interfaces are areas where rapid prototyping can be most effective. "Before the full system or even the full human–computer interface is designed, prototypes [i.e., whether of the system or of subsystems] can be developed and demonstrated with potential users to solicit early feedback on preferences of information representation" (Kossiakoff et al, 2011).

Risks that require some prototyping of the newly designed system may arise from a number of states and conditions surrounding infrastructure delivery:

- Emergent properties and/or complex interfaces apply
- New (hardware/software) components and elements
- Unusually high and/or tight performance requirements
- New or innovative materials, equipment, and processes
- Extreme environmental conditions (e.g., extreme weather)

Sir James Dyson (a *natural* engineer) tried 5127 prototypes before he could get his "cyclonic" vacuum cleaner right. Great engineers shall need far less!

Whereas prototyping has traditionally been used only for "mechanical" structures such as aircrafts, heavy machines and new inventions, there is nothing precluding the same technique from being used effectively in large and complex construction projects such as high-rise buildings, mega-malls, and complex/long-span bridges.

Since LIPs (i.e., skyscrapers, road interchanges, sea or air ports, long-span bridges) are growing in scale and complexity, they are now increasingly exposed to the conditions listed above (e.g., novel/innovative technology, complex interfaces, high performance requirements, extreme weathers, and above all else, emergent properties). The need has, thus, arisen to provide prototypes of such infrastructures to probe their "emergent behaviours".

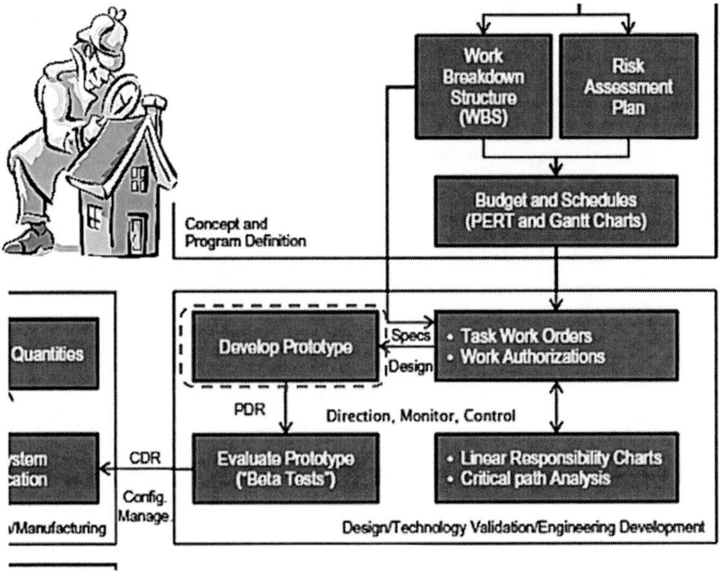

Figure 10.1 – Design Process Showing the Prototype (Adapted: Kossiakoff, 2011)

The Canton Tower (China), for instance, was designed as a multipurpose 604 m tall observation tower, the world's tallest up to 2010. The tower's height and slenderness have rendered it vulnerable to sustained winds of between 180 and 280 km per hour. These winds could cause the building to sway and make its occupants from the 90th floor up to feel motion sickness (due to a force of 1% of one's body weight).

It was expected that the rounding (at its corners) and twisting of the building might reduce the wind's adverse effects by up to 24%; but the question was, "By how much should the building be twisted"? A dedicated team of wind-tunnel experts prototyped a few models of the building and found that though a 180° twist was ideal, a 120° twist was far more practical.

Beyond these intricate issues of prototyping, Design Processes should also (and through necessity) support and/or enable risk treatment activities, including how to exploit opportunities or mitigate threats as relevant.

10.2 Engineering and Procurement Processes

Opportunities for the supply of goods and services available in the (local) market need to be *optimally* exploited, while market risks are at the same time being managed. Procurement Requirements are defined on the basis of the design and outcomes of the Engineering Process based on this principle:

> In terms of content, it makes no difference whether the contracting authority or rather the contractor carries out the processes during the application of Systems Engineering. This will make it feasible to involve the marketplace at any desired point of the system or asset life cycle. (ProRail, 2013)

Therefore, "During the identification of the procurement needs, the owner (or the contracting authority on their behalf) determines which life cycle phases of which objects (e.g., subsystems, components) are to be produced by market players (i.e., contractors, suppliers, advisors)." (ProRail, 2013)

"In the context of the overall agreement, procurement specifications should clearly state the criteria by which the acquirer [or the owner] will accept delivery from the supplier [or the contractor]. A verification matrix can be used to clarify these [acceptance] criteria." (Fossnes and Forsberg, 2006) This stance is recommended by INCOSE for Large Infrastructure Projects.

Adequate arrangements are required to manage the procurement process. "The engineering technical team [including designers] plays a key role in [the] development and evaluation of contract documentation [...] Depending on the type of procurement, a technical team member involved in source selection [i.e., identification of candidates] could be consumed nearly full time" (Systems Engineering Handbook, 2007).

Thus, the owner will specify the "scope and requirements" to the contractor. Except for accountability, everything else (i.e., activities and deliverables) beyond the issuing of Owner Requirements Specifications (ORS) falls into the contractor's hands; everything up to the issuing of ORS will, thus, fall into the owner's hands. In practical terms, the ORS is part of the "contract".

It shall not always be assumed that the ORS will be issued at the start of the project; it is perfectly appropriate to issue it at a later stage. Accordingly, the contractor is responsible for the consequences of decisions taken during the "*implementation of the ORS*" as issued through the procurement process.

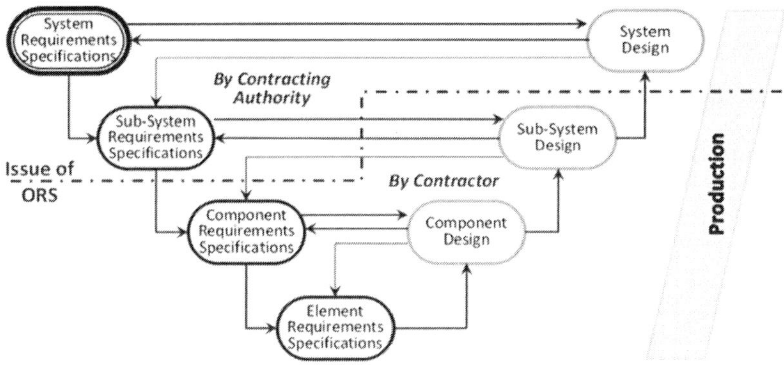

Figure 10.2 – Vee Model Showing Contracting Line (as per ProRail, 2013)

Figure 10.2 indicates that the owner may elect to issue the ORS at any end-of-stage of project delivery, but not in-between. The owner's organisation ought to secure the requisite capabilities for an effective delivery of any activities and deliverables preceding the issuing of ORS. Figure 10.2 also indicates the same ORS-line crossing the "right leg" of the Vee Model at some point. This is at the handover point where the owner, conversely, takes over the responsibility from the contractor. Sadly, the owner is often neither committed nor equipped to execute this last, but crucial life cycle step (as well as the Validation process that shall take place at this specific junction). Worse still, the owner may even lack the "readiness" to operate the system.

The engineers involved in the design process ought to also get involved in procurement. "When the professional [design engineer] is responsible for recommending contractors, he may be under a [judicious] duty to make reasonable enquiries as to their solvency and suitability for the work [...] In order to avoid liability for the cost of a contract that goes badly wrong, the architect [in the case of LIPs, the engineer] should investigate whether the contractors have [the] resources and skills necessary to execute the proposed works." (Hussin and Omran, 2009)

The foregoing provision will result, rightly so, in the engineering team sharing in the responsibility of the contractor(s) to effectively deliver the system. The engineering team should, thus, supervise the contractors' work:

> "Professionals owe their clients [i.e., owners] a duty to supervise or inspect the works with a view to ensuring that they are carried out to the standard contracted for. Reasonable supervision has been defined as such supervision as would enable the architect [or engineer] to give an honest certificate that the work had been executed according to the contract." (Hussin and Omran, 2009)

One practical way of attaining reasonable supervision is to ask the question:

> "[…] 'How will we know the proposed solution system remedies the undesirable situation or meets our needs? [viz., a Validation question.]' The ideas generated as responses to this [crucial] question lead to the quantitative acceptance criteria for the solution system and ways to measure the degree of remediation." (Kasser, 2015)

10.3 Engineering and Project Monitoring and Control

An effective and proactive PM approach ensures that managers have insight into the status of quality (i.e., outcome of engineering efforts), money, and time at any point of the life cycle. A project manager must be able to exhibit the results (i.e., "value") produced by each project activity (e.g., Design, Production), as well as the "cost and time" it took to complete the activity. Conversely, it is not practicable to *accurately* determine "cost and time" until most engineering activities/outputs (viz., designs pertaining to the "What to build" and "How to build") are adequately planned, agreed upon.

In the same vein, Jamshidi (2009) remarks that, "In their Framework for Assessing the Acquisition Function at Federal Agencies, the GAO [i.e., US General Accountant Officer] [...] set forth four cornerstones [for success] – [1] organizational alignment and leadership, [2] policies and processes, [3] human capital, and [4] knowledge and information management. Clearly defining and integrating [project] roles and responsibilities, effective communication and continuous [quality] improvement, empowering cross-functional teams, and monitoring and providing oversight to achieve desired outcomes [i.e., goals] were identified among the critical success factors".

It flows from the foregoing discussions and considerations that Monitoring and Control of activities and/or outputs is indeed a contributing factor to the success of Large Infrastructure Projects (LIPs).

In fact, Jamshidi (2009) argues that Monitoring is the quality and/or judicial function responsible for seeing that the "engineering goods" are delivered in accordance with [...] architectural standards, build-to requirements, and quality objectives. Kossiakoff supports this stance: "The project manager or PM, often with a technical or business background, is responsible for interfacing with the customer and for defining the work, developing the plans, monitoring and controlling the project progress, and delivering the finished output to the customer" (Kossiakoff, 2011). This active *monitoring* function also covers Operation and Maintenance, provided that project teams have carried out sections of Engineering Planning (see Chapter 09).

For this specific reason, "System Readiness" and "Operational Reliability" requirements should relate directly to the manner in which the system is to be maintained during its operating life. This demands that the system be designed to provide access for monitoring, testing, and repair requirements that are frequently not obvious at the project outset, but nonetheless must be addressed early in the development process (Kossiakoff et al, 2011).

Project Monitoring and Control or "Project Controls" shall ensure that time, money and quality (i.e., technical work) are controlled throughout the life cycle. "You can't manage cost [i.e., money] and schedule without managing the 'technical content'. It's where the time and money go" (Forsberg et al, 2005)—Many project managers are desperately trying to control cost and schedule way after they have lost control of technical works. Overruns are not the problem; they reflect the issues in 'technical content' of the project.

Project Controls heavily rely upon the WBS, which must be *dynamically* linked to project Scope and Configuration Management (see Figure 10.3). Otherwise, Earned Value Management can never answer the US$ 1 million question as to, "*How much of the project has been completed*?"—nor be able to identify potential technical problems in near real time (Kasser, 2015b). For "work performed" is not always equal to *progress* (in fact, there is no progress until the requirements are met to the satisfaction of stakeholders).

Project managers ought to monitor and control "*at what rate*", (1) work is performed, and (2) both time and money are being expended on the project. However, should "work performed" on a project not result in "progress" (i.e., as works are failing to meet the requirements), Monitoring and Control will get "jumbled" (viz., Earned Value Management will get disorientated) and project managers will still face or drive the *merry-go-round* of blame.

The Berlin-Brandenburg International Airport is a classic case of a failure of Engineering Management. The fire safety work packages were so flawed the project came to a near-standstill, with colossal cost overruns of 3+ billion Euros in 2012 (Deutsche Welle, 2013; Russia Today, 2013; Schetze, 2020).

Thereafter, the project leadership was no longer willing or able to forecast by when exactly the "Neverport" (as it was dubbed) might be completed. The opening date was postponed four times (i.e., 3rd June 2012; 13th March 2013; 27th October 2013; and even 17th of "*whenever*") before the Mayor of Berlin, who doubled as the "project manager", postponed it *indefinitely* (Deutsche Welle, 2013). Nonetheless, the latest forecasts were hinting at the 31st October 2020 as the opening date (Schetze, 2020).

The project management team, including the then Mayor of Berlin, was blamed for their conspicuous incompetence and their "lack of technical expertise". Politicians are not engineers and for ever replying, "*No comment ... I will consult the engineers*" to queries does not inspire much confidence.

However, the real incompetence and political ineptitude of the project team (led by Mr. Mayor himself) was to appoint some politically connected but fake fellow—an imposter—as the fire safety engineer. He had no other credentials than a business card that read, "Mr Z, Fire Safety Engineer". This is a valuable lesson for LIP leaders/managers under pressure to take instructions from corrupt politicians!

This improper, unlawful, and probably irregular (i.e., not having followed due process) appointment of a fire safety engineer resulted in the "failure of engineering" on the fire safety system (i.e., work package). For instance, the cable-ducts were too small; worse still, the blueprints were either faulty or missing. "*There are faults throughout the entire [new terminal] building*", groaned a public representative who had visited the building site at the time (Russia Today, 2013). It was projected that cost overruns may rise by ±150 million Euros per year of delayed completion, and operations might also be crippled. Owing to these "technical" inadvertences (e.g., appointing a bogus engineer, Mr Mayor himself doubling as the project manager, unsatisfactory works), Germany's third largest airport failed to open in 2010 as scheduled.

This predicament is certainly not peculiar to Germany, nor to their airports. All over the world, Large Infrastructure Projects are suffering from similar maladies; most LIPs are plagued with massive overruns. Like in Berlin, the fiasco might arise from a failure of Engineering. The increasing complexity in LIPs is surely making "engineering inadvertences" more likely than ever.

In addition to the above "Monitoring and Control" considerations regarding the delivery of large and complex projects (LIPs), the Dutch Ministry of Infrastructure and Environment (MIE) maintains the following guidelines:

I. An important aspect with regards to Monitoring and Control (i.e., Project Controls) is that the *independent* verification of these control aspects is carried out separately by third parties (albeit internal) and is not an integral component of the activities allotted to executing parties (ProRail, 2013).

II. The intensity and/or frequency level of controls in the Design and Development Process (i.e., EDD) is only justified for projects in which it is critical that changes and version control are properly maintained. The complexity of LIP-related issues and the need for explicit working methods is increasing the need for configuration management (ProRail, 2013).

III. Configuration should be defined as: "The functional and physical properties of a product [or system] as described in the technical documentation [e.g., specifications, drawings] and embedded in the product" (ProRail, 2013). Configuration, hence, ought to reflect the *content* and *scope* of the project.

IV. The failure to properly manage Project Support creates risk. For instance, the departure of key personnel, a dependence on software and a failure to adhere to safety directives. These factors alone can endanger the continuity of a large project. A link between PM and Controls (i.e., PM Office) clearly identifies the relationship between, on one hand, the project and people, and resources and processes on the other hand—which makes these relationships more manageable. Figure 10.3 below indicates such linkages (ProRail, 2013).

In fact, the left-hand part of Figure 10.3 (i.e., Specifications' connections between RBS, FBS, and SBS) was put into practice in Figure 2.5 above. To gain a further insight into the "relationships of the breakdowns" illustrated in Figure 10.3, one might need to recall SE Law number #9 of effective Systems Engineering as enunciated by Zane Scott (Figure 10.4):

> "Every system design involves three systems: The system being designed [i.e., Solution System], the system it will 'live' in [i.e., Context-System or operational environment], and the system used by the team to design it [i.e., Realisation-System or project environment]" (Scott, 2012; Mabelo, 2020b).

Indeed, the above system domains were referred to in Chapter 09 as follows:

(1) The **Development Programme/Project Planning & Control**
Realisation-System, the system used to design the solution.

(2) The **Systems Engineering Process**
Solution System, the system being designed and constructed.

(3) The **Engineering Specialty Integration**
Context-System, the environment where the system shall live.

Finally, Monitoring and Controls is not an "isolated" domain. At the Design stage, system requirements are converted into functions to be translated into system components. During the subsequent Construction stage, the scope of work involved in materialising the system is translated into appropriate and *interconnected* work packages, the implementation of which should be "monitored and controlled" by way of periodic and *ad hoc* Progress Reports.

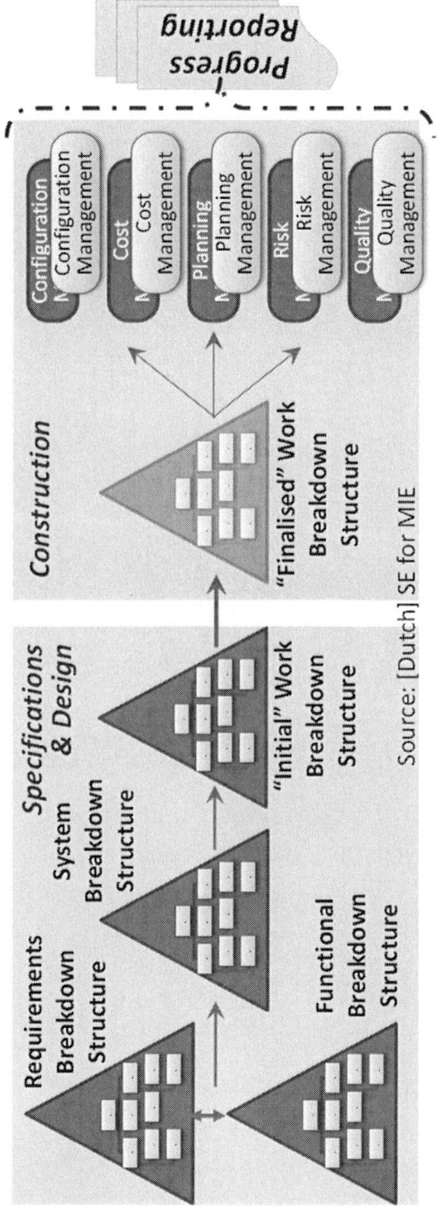

Figure 10.3 – Relationships of Breakdowns (Adapted: ProRail, 2013)

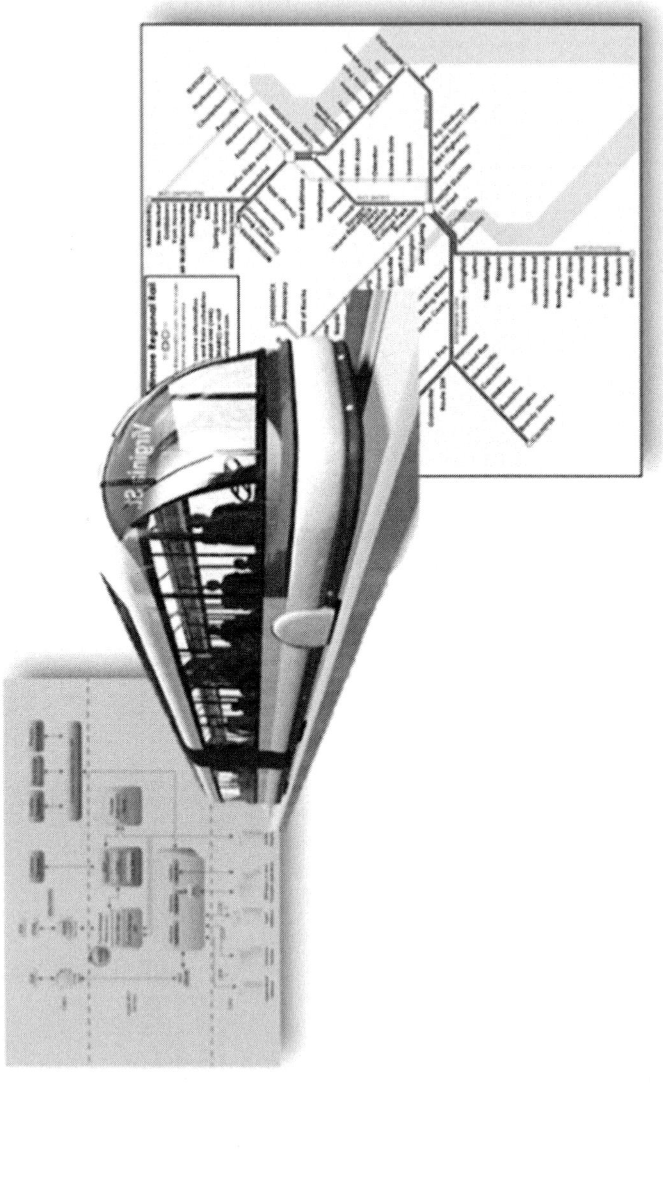

Figure 10.4 – Realisation-Solution-Context Systems (as per Scott, 2012)

CHAPTER 11

THE DESIGN THINKING PROCESS

Understanding the concept of Engineering Design as discussed hitherto has definitely proved useful to EDD; similarly useful is "*grasping the essence*" of conventional, best practices, and innovative/advanced design processes. However, even as Scott has put it, "To catch [i.e., to design] a system you have to think like one"—a law of effective SE (Scott, 2012). Therefore, this chapter shall seek to introduce the systems way of "Thinking about Design".

Dam and Yu Siang define Design Thinking as a systemic tool, as a design methodology that provides a solution-based approach to solving problems. "It is extremely useful in tackling complex problems that are ill-defined or unknown [e.g., complex infrastructures], by understanding the human needs involved, by reframing the problem in human-centric ways, by creating many ideas in brainstorming sessions, and by adopting a hands-on approach in prototyping and testing"—five crucial stages (Dam and Yu Siang, 2020).

These five Design Thinking stages constitute the "core" of this systemic methodology; "Understanding these five stages of Design Thinking will empower anyone to apply the Design Thinking methods in order to solve complex problems that occur around [most of] us—in our companies, in our countries, and even on the scale of our planet" (Dam and Yu Siang, 2020). This design consideration applies to Large Infrastructure Projects (LIPs).

Design Thinking (DT) is therefore a methodology used by *designers* to solve complex problems, and find desirable solutions for clients. It is not problem-focused, but rather it focuses on solutions and actions that are oriented towards creating a preferred future, drawing upon logic, imagination, intuition, and system reasoning, to explore possibilities of "What could be", and to create desired outcomes (as per the ConOps) that benefit users/customers. Many prominent DT practitioners have subscribed to this perspective:

- Nigel Cross, author of '*Designerly Ways of Knowing*'— "Everything we have around us has been designed. Design ability is, in fact, one of the three fundamental dimensions of human intelligence. Design, science, and art form an 'and' not an 'or' relationship to create incredible human cognitive ability." (Cross, 2006)

- Tim Brown, CEO of IDEO — "Design Thinking can be described as a discipline that uses the designer's sensibility [i.e., empathy] and methods to match people's needs with what is technologically feasible and what a viable business strategy can convert into customer value and market opportunities." (Brown, 2008)
- Roger Martin, author of '*Design of Business*' — "Design-thinking firms [and project teams] stand apart in their willingness to engage in the task of continuously redesigning their business […] to create advances in […] innovation and efficiency—the combination that produces the most powerful competitive edge." (Martin, 2009)
- L.C. Hwa, author of '*Design Thinking: The Guidebook*'— "Design Thinking believes (and has been proven) that the people who face [multi-faceted engineering] problems are the ones who hold the key to their problem's answer. Design Thinking is a human-centered problem-solving tool [...] which devises feasible and viable Big Idea solutions. The key to the process is empathizing with the users (citizens) to uncover unmet needs by understanding their beliefs, values, motivations, behaviors, pains, gains and challenges and to provide innovative solution concepts." (Hwa et al, 2020)

The UK Design Council similarly promotes a Design Thinking framework that maps out four distinct phases, illustrating 'divergent' and 'convergent' stages of the process; namely, (1) Discover, (2) Define, (3) Develop, and (4) Deliver (see Figure 11.1 below).

However, the five-stage DT model proposed by the Hasso-Plattner Institute of Design at Stanford (d.school) seems to have gained renown in the industry. Authors such as Dam and Teo claim that, "d.school is the leading university when it comes to teaching Design Thinking" (Dam and Yu Siang, 2020).

Indeed, Stanford University, where DT might have originated, recommends the five-phase process as a method of solving complex problems, including Engineering Design (as applicable to large and complex infrastructures, LIPs). The five stages of Design Thinking, according to d.school are as follows: (1) Empathise, (2) Define (the problem in human-centered terms), (3) Ideate, (4) Prototype and (5) Test (Dam and Yu Siang, 2020).

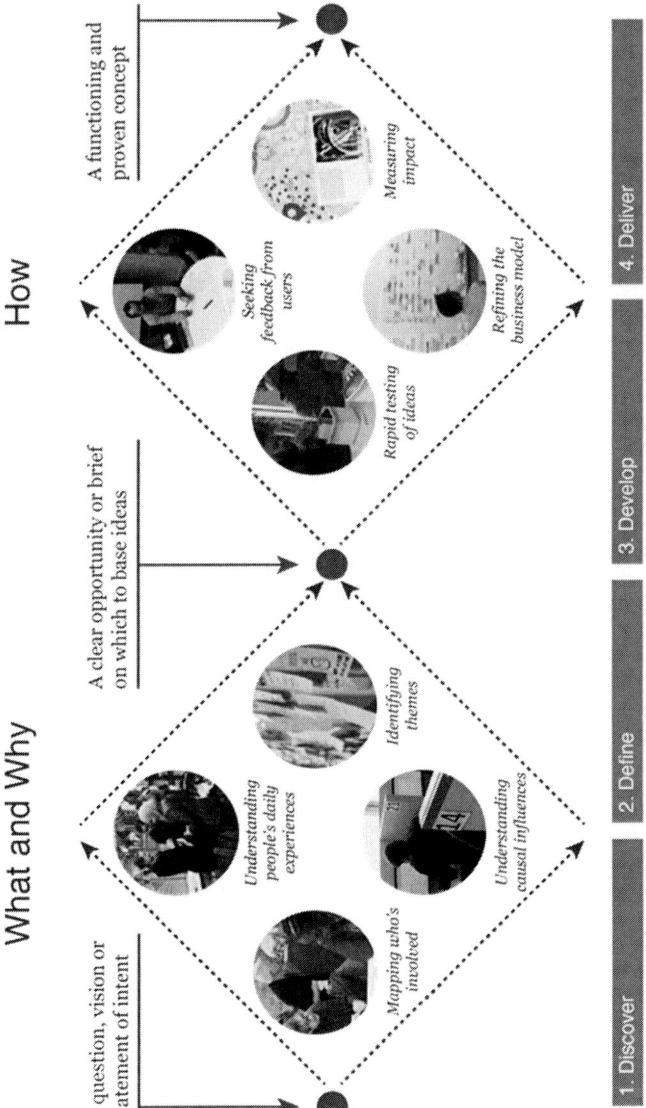

Figure 11.1 – Stages of Design Thinking (Source: www.CreativityWork.com)

Therefore, Figure 11.2 depicts the five essential stages of Design Thinking:

1. **Empathise:**

This first phase is about developing an understanding for the challenges, problems, and needs and/or requirements (i.e., problem understanding). In this phase, detailed research and on-site observations are carried out on the customer's needs or problems—including their concerns and predicaments.

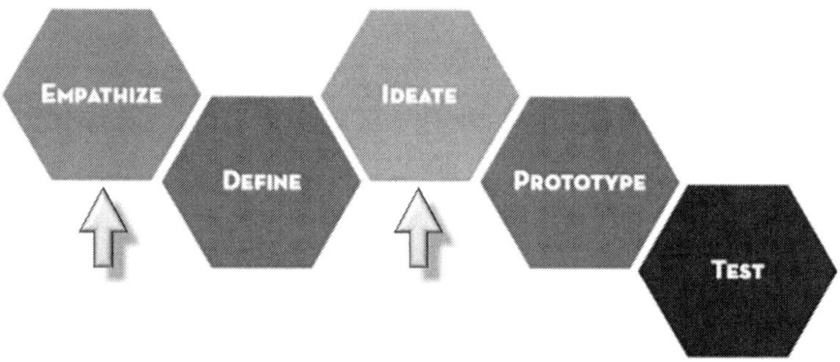

Figure 11.2 – Five Phases of Design Thinking (Adapted: Stanfordonline, 2013)

The main purpose of this phase is to gain an "empathic" understanding of the problem one is trying to solve. Empathy is crucial to a human-centred design process such as DT, and empathy allows Design Thinkers to set aside their own flawed assumptions (or prejudice) about the world and seek to gain insight into the users/customers and their needs.

This involves consulting subject matter experts (e.g., anthropologists, sociologists) to find out more about the area of concern through observing, engaging and empathising with people to understand their experiences and motivations, as well as immersing oneself in the physical environment so as to gain a deeper personal understanding of the issues involved (Dam and Yu Siang, 2020); in the context of LIPs, the author will refer to "*Putting Empathy into Operations*" (Mabelo, 2020b).

2. **Define:**

During this stage, the information generated and gathered during the previous "Empathise Stage" is collated in order to elucidate the problems. This is where the observations are analysed and synthesised in order to define the "core" problems and other relevant and/or pressing issues that have been identified hitherto; a wrong framing of the problem statement would bedevil the ensuing steps of the Design Thinking process.

Therefore, the problem statement shall be defined in a human-centred manner. Furthermore, it is often recommended as good practice that after the observations, the findings should be condensed to a "persona" (a single prototypical user/customer) whose problems/needs and worries ought to be summarised in a clearly defined design question (Dam and Yu Siang, 2020).

3. **Ideate:**

From this stage onward, designers are ready to start generating "solution" ideas. Armed with an "empathic" understanding of the users involved and their needs in the "Empathise Stage" and on the strength of the human-centred problem statement, the DT team (i.e., system designers) shall start *thinking* "outside-the-box" in order to generate or identify new solutions to the problem statement as articulated earlier (i.e., problem framing). It is crucial that alternative ways of viewing the problem are considered.

It is only in this phase that the actual brainstorming process takes place; creativity techniques, as many as might be available, should be used. There is a plethora of Ideation techniques such as the (Active-) Brainstorming, DELPHI, What-If Scenario, Storytelling (i.e., User-Stories), and Interviews. Soon after their generation, "ideas" should be clustered by intention to identify themes. This clustering helps convert ideas into design concepts.

Strictly separated from this, the ideas or concepts can then be analysed in a customer-oriented manner in an attempt to identify weak points, and a selection-decision can be made on the basis of an idea/concept evaluation.

4. **Prototype:**

Selected ideas must be visualised as quickly as possible, made tangible, sketched, designed, modelled/simulated, etc. Following the technical field, one can speak here of "Rapid Prototyping", in which the prototype development applies definitely not only to products, but also to services. While a prototype is basically meant for visualising a potential solution, there is nothing precluding the designers from applying a Detailed Design process and, thus, benefitting from the "realism in design" it can afford.

Thus, the design team should produce a number of inexpensive, scaled down versions of the product/service or specific features found within the product/service, so they can investigate the problem solutions generated in the previous stage. Prototypes may be shared, tested within the design team itself, in other departments, or on a small panel of experts outside the design team. This is an experimental phase, and the aim is to identify the best possible solution for each of the problems identified during the previous three stages. The solutions are implemented within the prototypes, and, one by one, they are investigated and either accepted, improved and re-examined, or rejected on the basis of the users' *real-life* experiences (Dam and Yu Siang, 2020).

5. **Test:**

In this final phase, the ideas (further translated into design concepts) are to be developed and tested through further experiments and customer feedback. In addition, important development, production and market issues have to be clarified, taking into account the customers' responses in terms of articulated and latent needs—that are harder to find out and define, but they are important sources of inspiration for innovation (Hwa et al, 2020).

Designers or evaluators should rigorously test (i.e., the look and feel of) the complete product (e.g., visual proxy) using the best solutions identified during the prior "Prototype Stage". This is an iterative process; the results generated during the testing phase are often used to redefine one or more problems and inform or increase the understanding of the users, the conditions of use, how people think, behave, and feel, and to empathise. Alterations and refinements are made in order to rule out problems in solutions and derive a deep understanding of the product and its users.

Of these five steps, "Empathise" (i.e., problem-finding approach) and "Ideate" (e.g., Brainstorming of potential solutions) are the most prominent phases of this dynamic process. However, a Design Thinking paradigm (which shall consist of "Bias towards Action", "Mindfulness of Process", "Radical Collaboration", "Reframing", and "Curiosity") is required for the process to prove effective:

(1) The Empathise Phase:

Many frameworks could be used to gain empathy and classify observations in insightful ways—allowing designers to establish "*How customers or users feel about the problem at hand (and the prospect of a solution)*". Two of the most commonly used frameworks, namely, the Empathy Map (see Figure 13.3) and Maslow's Hierarchy of Needs (see Figure 13.4) are briefly discussed here.

Since what people "say and do" is actually a result of what they "think", designers are expected to interview customers about their deep-rooted needs/issues or desires:

> "The most secure source of new ideas that have true competitive advantage, and hence, higher margins, is customers' unarticulated [or hidden] needs." (Liedka, 2014)

Design Thinkers must *unearth* such unarticulated or emotional needs out of users and customers, probing what they "think" and "feel", not just what they might say. This approach is similar to the "*diving under the tip of the iceberg*" (which was illustrated in Figure 2.4) to establish the actual needs.

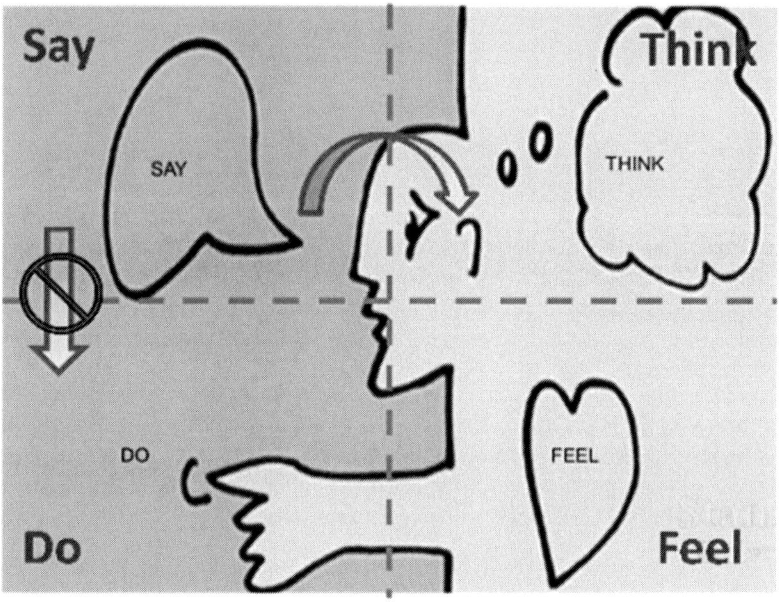

Figure 11.3 – Empathy Map (Adapted: Stanfordonline, 2013)

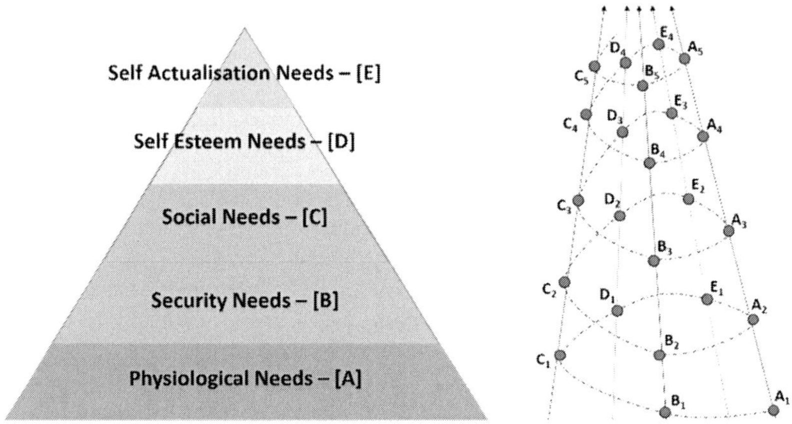

Figure 11.4 – Dynamic Version of Hierarchy of Needs (By the author)

According to Maslow (Hellriegel and Slocum, 1996), infrastructure users shall have their needs met at the lower levels of the pyramid (i.e., water, transport, energy, health) before they are *fruitfully* motivated to tackle and/or request the next levels (e.g., aesthetic designs). Thus, sub-Saharan nations ought to provide more *essential* infrastructure than skyscrapers.

It would of course hurt national pride that a country of 183 million people might not boast of as many skyscrapers as other developing nations on the continent. But does competing on that front really make sense when only 48 percent of your population has access to basic drinking water sources, and worse still, only 33 percent have access to sanitation? Maslow will disagree!

In terms of the above concerns, Figure 11.4 involves dynamic perspectives. The author's rendition of the pyramid (as a "converging spiral") adds a *dynamic* dimension to Maslow's theory to align it to life's reality as follows:

i. All five types of "needs" (ranging from A to E) are *always* present in a person's life
ii. The focus or emotional burden is on a *single, specific type* at any point in time
iii. The focus or burden keeps shifting *upwards* along Maslow's "spiral of growth"
iv. The utility-gap between types of needs (e.g., moving from A to B) *contracts* (i.e., takes less time/effort) in upper levels of the pyramid

(2) The Ideate Phase:

This is the phase where the rubber of innovation hits the road of problem-solving. For it is in generating "innovative" ideas that novel [engineering] solutions would be found. Thus, "We moved from thinking of ourselves as designers to thinking of ourselves as design thinkers [which shall apply to design engineers]. We have a methodology that enables us to come up with a solution that nobody has before", notes David Kelley (Tischler, 2009).

This is exactly what aspirant design engineers should learn. "Design Thinking is all about unlocking potential and creating creative confidence in our students [...] we believe the next generation of innovators and leaders [e.g., design engineers] need to be great design thinkers", Tom Kelley argues (High Resolution, 2017).

It is common cause that ideas will gush out of the brain, if given the right stimuli. Indeed, it does not take a lot of energy for the brain to deliver those good ideas. "Neural Darwinism posits that the brain has evolved through resource competition and adaptation to process [new] information at the lowest possible energy state" (Edelman, 1987). In fact, it needs less *energy* than it takes to switch on a 40-Watt household light bulb.

Imagination is all it takes. "A basic precondition for independent design is a *lively* imagination. Such imagination is required to do original work" (Haik and Shahin, 2011). However, flooding the proceedings with ideas is not necessarily a good thing. Some convergence is required to focus those ideas on the problem at hand. This calls for stopping oneself from choosing a solution from the *first few ideas*. Techniques such as Brainstorming and Mind Maps are used in a workshop setting to generate and focus on "good" ideas. Accordingly, Mary Lou Cook (a famous actress, singer and dancer—a creative person by vocation) had this much to say regarding "creativity":

> "Creativity [i.e., generating loads of novel and/or innovative ideas] is inventing, experimenting, growing, taking risks, breaking rules, making [and fixing] mistakes, and having fun [but not necessarily, "*singing and dancing*"]." (Zimmer et al, 2008)

Therefore, the working environment for an effective "Ideation" process is designed to promote creativity. One could allude to the three Ps of Design Thinking: People (i.e., human beings), Process (i.e., the problem-solving process) and Place (i.e., the working space) must be considered for the development of a successful idea. A fourth P can be Partnerships, since a large number of partners (i.e., workshop participants) ought to be involved in the development and implementation of ideas (Mueller-Roterberg, 2018).

Most brainstorming techniques fit the part. The effective brainstorming process should involve four distinct stages as follows:

(1) Problem Framing – is about stating the problem in an open-ended mode, by setting goals, not the method or means of solving such a problem
(2) Warming Up – is about lifting participants from an *analytic/critical* paradigm to the required *synthesis/non-judgemental* paradigm
(3) Idea Generation – is about facilitating the gathering of ideas, by deferring judgement and following *headlines* or design concepts
(4) Ideas Grouping and Selection – entail taking time to group, select, and process ideas for action (Figure 11.5), otherwise nothing much will change thereafter

Unfortunately, common versions of brainstorming techniques are guilty of skipping three of the above stages; brainstorming sessions are often reduced to a single stage, namely Idea Generation. This poor attitude can expedite the process, but it might leave behind "*a dry and withered tree*", in terms of richness and implementation of ideas.

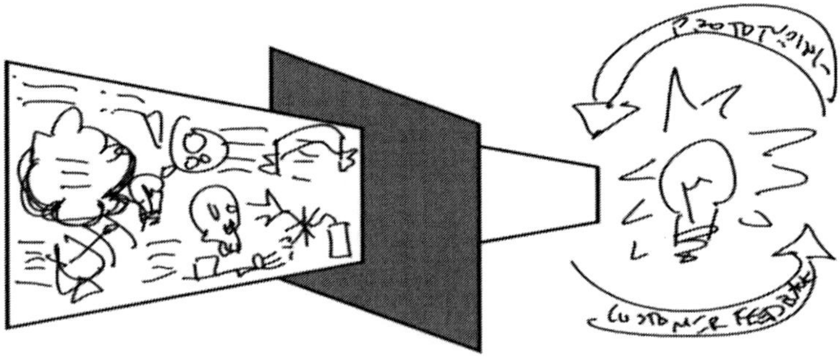

Figure 11.5 – Ideas Grouping and Selection for Action (As per Stanfordonline, 2013)

Therefore, Kasser (2015) promotes Active Brainstorming as a way to increase the number of ideas generated by *initially* brainstorming using the nine Holistic Thinking Perspectives (HTPs) and then combining them with six Kipling-type questions, i.e., "*Who, What, Where, When, Why, and How*".

Referring to DT as a systems way of "Thinking about Design", Mueller-Roterberg (2018) argues that, "The approach is ultimately applicable to all kinds of business [and engineering] ideas, whether they have a product or service character. The first 'mouse' for the Macintosh computer was created after a similar approach, or the first toothbrush with a wider ergonomic shaft". The process is an integrative approach: "This means that for problem-solving, the process of problem-solving [e.g., System Design] is considered together with its framework conditions" (Mueller-Roterberg, 2018).

Since DT is "*ultimately applicable to all kinds of business [as well as engineering] ideas*", it follows that this systemic approach to design is suitable and adequate for bringing innovation in the design of large and complex infrastructure-related systems such as those involved in most of Water, Energy, Transportation, Waste Disposal, Information Technologies, and Telecommunications (or "*WETWIT*" industry, as the author coined it).

By infusing Design Thinking throughout the Engineering and Development (EDD) process (which involves the design of both the "infrastructure system" itself, as well as the "process" of Construction), the designers will acquire the "*sensibility and methods to match people's needs with what is technologically feasible*"; for indeed, "*the next generation of innovators and leaders [i.e., design engineers] need to be great design thinkers*". Such a combination of design tools yields the most powerful competitive edge.

This power arises when designers would refrain from "*choosing the first good ideas*". For instance, Daylight™ was given the challenge of getting kids in America to move more and, thus, fight child obesity. The project started off with a "good idea": Provide kids with digital music players fitted with motion sensors and apportion rewards based on their activities (e.g., physical motions). However, by talking to kids about their needs, likes, dislikes, and worries, the team began to notice patterns that led to design principles (e.g., "*Facilitate social interaction at all times*"), which were used to create some prototypes. A small monitor to clip on clothes or a pocket linked to an online portal where kids could celebrate their success was chosen—56% upturn in activities ensued in 6 months (Daylight, 2014).

Further, the architect at a certain university elected to "design" a centralised storage of gases for the chemical lab. Using a 12-pipe copper reticulation, she then *circulated* the various gases throughout the whole facility. After a period of dysfunctional and unsafe operations, intense consultations and "heated" debates ensued with the users (e.g., lab operators), which led to the adoption of a *redesign* approach and criteria. The new engineering team conducted a few tests and validation (i.e., before acceptance) with the lab personnel, and proceeded to change the design of the gas supply to the satisfaction of all concerned. Since then, lab operations have been running far smoother at the facility.

"*A secure storage of the right gas canister close enough to the place of use and an appropriate piping to connect the two points*", is all the design team should have devised in the first place. But the architect had "disrespected", or in other words, failed to consult the lab users and she ended up with some *ill-suited* design. In this case, DT techniques and paradigms could have prevented the flawed design. The total costs of redesign and retrofitting amounted to more than US$ 12 million, not to mention the disruptions to lab operations, because the initial project team failed to exercise empathy.

Hence, the author has devoted an entire section to the notion of "*Putting Empathy in Operations*" in a previous publication (Mabelo, 2020b). He advocates the "planting of one's heart" (i.e., empathy) in operations before the "applying of one's mind" to engineering designs and the development life cycle. For it is the problems, needs, issues, challenges, ills and wills in operations (i.e., as experienced by consumers, users, operators, and broader communities) that ought to be taken into account in devising the new Solution System. Forsberg maintains a "system's view" to that effect:

> "The most popular models in the development project environment focus either on project administration, technical development, or process improvement, often to the exclusion of the other areas. All too often projects proceed with innovation and sophisticated development without paying heed to the evolving business case [e.g., improved operations, benefits to customers and to other stakeholders]. Furthermore, the managers of supporting subsystems or items usually have little knowledge of the driving business case and of the derivative cases at their level. This lack of awareness of the underlying business issues [i.e., the "*voice of the customers*", their unarticulated needs] stems from inadequate collaboration between the business and technical disciplines and can lead to a wrong project solution [i.e., products/services unfit for operations] that is ultimately rejected by the users, customer, or marketplace." (Forsberg et al, 2005)

The question then arises as to how to essentially achieve this "integration in design" that enables design engineers to develop new products and services that meet customer needs—the right solution the first time? Forsberg still has this practical advice to give:

> "How then to best accommodate the evolving business case and to have it drive the technical and financial decisions throughout the project life cycle? The answer begins with internalizing the integrated process model presented here, tailoring the processes, and then putting the practices to work. The set of models presented in these pages builds on the natural synergies of project management and systems engineering [...] Installing an integrated project management and systems engineering culture, based on the models in this book, coupled with training and certifying key team members has significantly improved project success rates. Moving beyond success to a strong project culture and a predictable performance improvement program [e.g., a company-wide adoption of Design Thinking (DT)] can represent a distinct competitive advantage." (Forsberg et al, 2005)

It is as if Forsberg was talking *in premonition* about this specific book and its approach. Indeed, the delivery of large and complex infrastructures would benefit from applying Design Thinking. On that note, it is greatly encouraging that Victoria State (Australia) has adopted a Design Thinking approach (i.e., The "*Healthy Together Victoria*") to design and improve the infrastructure where their population should "*Live, Work, Learn, and Play*" (Victoria Government, 2014). It is all the more encouraging that some tertiary institutions are nowadays exposing engineering students to Design Thinking.

CHAPTER 12

PRACTICAL APPLICATIONS

Every given system is part of a larger entity. At times, this larger entity can be classified as a separate (meta-) system in itself (rather than simply the environment, or "nature"). These situations are alluded to as "SoS" and tend to exhibit seven distinct characteristics:

> "(1) operational independence of individual systems, (2) managerial independence of individual systems, (3) geographic distribution, (4) emergent behaviour, (5) evolutionary development, (6) self-organisation, and (7) adaptation." (Glass et al, 2008)

Most mega projects [e.g., LIPs] concern themselves with large and complex infrastructures that are construed as SoS. Therefore, "The design engineer must learn to think independently, to draw conclusions, and to combine solutions. Many believe that they can acquire this skill set by attending lectures and reading textbooks. They fail to realize that they are only accumulating one fresh item of knowledge after another" (Haik and Shahin, 2011).

Understanding, logical deduction, and judgement cannot be conferred from the outside; on the contrary, they are better acquired by diligent thinking and working with the knowledge already attained (Haik and Shahin, 2011). For "Mastery of theory, along with mastery of practical skills in the field, is a hallmark of professionals" (Koskela and Howell, 2002). The author, thus, maintains that practical application of the theory is crucial to mastery:

> "Therefore, LIP practitioners will require as many opportunities as their circumstances might dictate to master the crucial skills of managing engineering processes; part of which is systematic (i.e., process), and part of which is artistic (i.e., human-centred). This mastery would not happen automatically, but it will require commitment and lot of effort." (The author)

This "practical" section, which would not claim to be exhaustive, presents the reader with an opportunity to muster and sharpen their own practical skills as far as design and development of infrastructure systems (which are often SoS) are concerned. Some of the practical exercises included herein might require a measure of engineering mastery, which would perhaps prove challenging to non-engineers. In such taxing instances, the reader is kindly instructed to simply skip those specific exercises, except when they are willing to enlist support from experts.

However, "Successful professionals believe in the necessity of self-improvement. They know that if they do not continually invest in their own learning and development, they will become obsolete" (Jamshidi, 2009). Like in any other serious endeavour, "The journey to [Engineering Design and Development (EDD)] mastery might be long and taxing—yet practice shall make perfect" (Mabelo, 2020b).

In particular, engineering practitioners involved in the design and development of large and complex infrastructures will benefit from trying out different and innovative ways of thinking about design. In so doing, they will break free from the traditional design mould that might have accustomed them to "*rushing into detailed design*" or, worse still, to a counterproductive pattern of "*design-by-force-of-habit*". It thus calls for the adoption of the "Spirit of Design" and, furthermore, for shunning pernicious design "inadvertences" that still compromise and bedevil the LIPs industry.

12.1 The Spirit of Design

Engineering Design dates back to ~2700 BC when the step-pyramid of Pharaoh Zoser was built by Imhotep—the first *engineer* known by name in human history, apart from those in the Bible who built Noah's Ark and the Babel Tower. A lot has happened since and engineering as such has evolved and has proceeded from an *esoteric guild* to crafts to chartered professions.

According to its current meaning, "Design [i.e., a product of engineering] is a drawing or set of drawings showing how a building or product [or system] is to be made and how it will work and look" (San Cristobal et al, 2018). It is also "*the process of devising a system, component or process to meet desired needs*". Hence, "The concept of Design Process is the description of the sequence of steps taken by a designer [i.e., engineer] in converting an abstract set of requirements into the definition of a physically realizable product or system" (Sivaloganathan and Al-Marzouqi, 2014).

Furthermore, allusions were made from the onset about similarities and differences between artistic design (e.g., sculpture, painting, weaving, music, poetry) and engineering design. To ascertain the uniformities and dissimilarities between artistic and engineering endeavours, the author discussed the "design approach" with two Johannesburg artists (one being the renowned painter, Mr Ricky Burnett) and a veteran design engineer (Mr Bheki Mpala).

The outcomes of those structured interviews are recapped in Table 12.1 below. From their responses, the following inferences regarding their "attitudes" to design could be drawn:

a. Both artists and engineers appreciate the "innovation and creativity" that arise out of an open and inquiring mind; however, they pursue such through different avenues.
b. Artists are more inclined to new ideas than engineers might be (who stick to 'tried and tested'), though artists would not always actively seek novelty and innovation.
c. In the artist's world, ideas seem to rather "flux" into the final products, which is the ultimate object. Engineers generally entertain some artificial line between concepts and products, but this often only works "on paper"; on site, they too tend to blur that dividing line, behaving just like artists.
d. While artists are inspired in their work by their own ideas, with little or no concern regarding clients' wishes, whims and other desiderata, engineers believe in accommodating requests so long as they fit the project. Consequently, artists gauge success based on their own criteria, whereas engineers generally consider delivery requirements.
e. For artists and engineers alike, the more they gauge success by the client's criteria, the more they tend to estimate and, at times, control the expenditure of resources.

The foregoing considerations and outlooks attest to the *universality* of the Design Process. As innately useful as it might be, design mostly gains potency when upheld by innovation. For instance, Robin analysed case studies from two prominent designers and concluded:

> "The basic concept of both Dyson's innovations arose from a mental transfer of technology from one application to another. Sanders, on the other hand, tended to seek analogies between the problem he was trying to solve and products [or systems] or components with similar functions." (Roy, 1993)

<table>
<tr><td colspan="2">Q:01—Would you prefer your designs to resemble or differ from previous ones?</td></tr>
<tr><td>Artist 1</td><td>Depends on the development of the idea. While an idea runs its course, you stay with it, when it has run its course, you change; thus, change for the sake of novelty is pointless.</td></tr>
<tr><td>Artist 2</td><td>I don't really mind if my ideas are totally new and different. I suppose there is often an unconscious link to previous works, and things connect in ways I might not be aware of, but I certainly enjoy new challenges.</td></tr>
<tr><td>Engineer</td><td>It is good keeping to previous designs that are tested and have stood the test of time. Nevertheless, in an evolving society, it is important to use innovative designs, new methods of building and types of material to help with developing unique designs that can be admired in the new era.</td></tr>
<tr><td colspan="2">Q:02—Do you make a clear difference between the idea/concept and the artefacts?</td></tr>
<tr><td>Artist 1</td><td>No, the meaning is embodied in the material. Until the material owns and embodies the idea, the result can at best only be illustrational.</td></tr>
<tr><td>Artist 2</td><td>Although deferent, the two aspects are intrinsically connected and the process influences the concept.</td></tr>
<tr><td>Engineer</td><td>Designs done in the office are usually a conceptual prototype of the end product which looks good on paper. However, in the real world, it brings about a couple of changes that are sensitive to the end user.</td></tr>
<tr><td colspan="2">Q:03—Would you prefer working based on your own ideas or requests from clients?</td></tr>
<tr><td>Artist 1</td><td>I never entertain any clients' ideas.</td></tr>
<tr><td>Artist 2</td><td>I generally prefer working based on my own ideas.</td></tr>
<tr><td>Engineer</td><td>It is important to accommodate the client's needs as he is the end user [...] both the designer and the client need to reach a satisfactory decision as to what works best, taking the aesthetics into consideration as well.</td></tr>
<tr><td colspan="2">Q:04—Do you work out the required resources or rather figure it out as work goes?</td></tr>
<tr><td>Artist 1</td><td>I find this question irrelevant to vocational drive that underlies my work.</td></tr>
<tr><td>Artist 2</td><td>I generally calculate time and resources before commencing a project.</td></tr>
<tr><td>Engineer</td><td>It is important to understand the related costs and resources required.</td></tr>
</table>

Q:05—*How do you gauge "success", by your own criteria or by client satisfaction?*	
Artist 1	Mostly by my own criteria, but applause is always nice.
Artist 2	If I am not pleased with the result, I will not even show it to the client. I need to be completely in charge of the quality of my work.
Engineer	I gauge success when the project is complete and has been commissioned with minimum issues, when it is completed within budget and on time.
Q:06—*What would you advise anyone who wishes to learn innovation or creativity?*	
Artist 1	Be curious about possible outcomes and work to express this curiosity.
Artist 2	Read a lot, collect interesting things, look at art, and learn to play.
Engineer	Get a mentor, develop your education. Once you are confident, having gained experience, then one can have an open mind and be innovative.

Table 12.1 – Summary of Interview Responses (Artistic vs. Engineering Designs)

Many eminent artists follow precisely the same approach to design works. Therefore, great engineering ought to blend art (i.e., creativity) and science (i.e., process). For instance, when he was designing the *Huayra*, Horacio Pagani (an engineer) was quoted as saying:

> "Our mission is to create a car that creates emotion [i.e., that relates to and/or satisfies the emotional needs of potential users] [...] to create that emotion means combining art and engineering." (PantherMovieProductions, 2014)

The *Huayra* was the result of an engineering crusade (i.e., engineering marvel), but also a work of art—a philosophy that evolved from his lifelong fascination with Leonardo da Vinci. "*Every part of the car is a work of art*", said a member of the technical team (PantherMovieProductions, 2014); the *splendid coupé* is so desirable it sells for over a million Euro (i.e., at the factory price). The "quality" of the car commands such an exorbitant price.

There is no doubt that humans (i.e., customers) are both rationale and emotional beings. As such, customers would be expected to make rational and emotional choices; in many instances, their decisions might turn out both rational and emotional *at the same time*. For this reason, and the author's rendition of Maslow's Pyramid of Needs, engineered products and/or services must address both the rational side (i.e., conformance to stated requirements) and the emotional side (i.e., unarticulated, yet pressing needs—*getting customers to want them!*). The latter endeavour often proves hard to achieve in practice, which makes it more competitive—affording large margins to the few that can attain it.

In fact, the American Society for Quality defines Quality as, "*The totality of features and characteristics of a product or service that bear on its ability to satisfy given needs*"—thus, it shall reflect conformance to specifications or customers' requirements. Quality is similarly about the inherent or distinguishing characteristic, a degree or grade of excellence when referring to a particular product and/or service—good/high quality. The above definition conveys the "objective" meaning of the word quality. However, there is a "subjective" meaning of the same word that now refers to "taste, preference"; "*Quality is about exceeding customer expectations*"—about "*WOW-ing the customer*", which can only prove *personal* as it depends and varies from one individual to the next.

It follows from the foregoing argument that the Engineering Design and Development (EDD) endeavour should seek to address not only the stated requirements, but also the visual appeal and the "emotional attraction" aspects of the infrastructure-system under consideration.

It is lamentable that many design engineers (in particular those involved in LIPs) would basically focus on meeting system requirements (i.e., "*What the system shall do*") to deliver the right functionalities (i.e., a system that is *fit-for-operations*), without much accommodation to the visual appeal and other aesthetic (i.e., artistic) design requisites. Here the author argues that no design is *thorough* unless both aspects are incorporated.

Indeed, when asked whether he would support the idea of teaching "artistic design" to engineering students to prepare them for applying a blended approach to design, Ricky Burnett (an eminent Johannesburg-based painter) had this to say:

> "I would indeed, especially insofar as the relationships of built things to their natural surrounds and to each other are surely of vital concern to the quality of our habitable world".

Ricky Burnett definitely backs the idea of combining technical and artistic inputs; further, he'd rather take a system's view of the "designed" world, by stressing "relationships" between the designed items and the environment.

Engineers should therefore begin to appreciate that artistic input always proves beneficial to infrastructure design. Henceforth, artists such as architects, sculptors, painters and other horticulturalists must no longer be treated as "intruders" by engineering teams, but as important partners to the design endeavours. They provide crucial inputs particularly when it comes to "*shapes and forms*" of elements—which enhances the visual appeal, likeability, and desirability of infrastructures, thus attracting high tariffs.

This stance will be consistent with the author's rendition of the Maslow Hierarchy of needs; satisfying basic infrastructural needs (e.g., supply of transport) ought not to preclude aesthetics for the benefit of operators/users.

The INCOSE Handbook on Systems Engineering (Fossnes and Forsberg, 2006), one of the most advanced and esteemed engineering manuals, refers to *visual appeal* as follows:

> "The methodology should rule out absurd or obviously non-optimal combinations of elements, and seek particularly <u>appealing</u> [i.e., visual] new combinations of the elements".
>
> "Functions that happen "all the time," such as safety, <u>aesthetic functions</u>, etc., are placed above the critical path functions".
>
> "Selection criteria [for systems] are the quantifiable consequences of system implementation and operation [...] Selection criteria include [among others]:
>
> 7. Measures of risk
> 8. Measures of <u>quality factors</u>
> 9. Measures of <u>subjective factors</u> that make the system more or less acceptable to customers, users, or clients (e. g., <u>aesthetic characteristics</u>)".
>
> [The underlining of phrases is added by the author for emphasis purposes]

In light of the above, one could only wonder where this false notion came from that "art" (which is the "science" of visual appeal) ought not to feature next to engineering. It is for a reason that Design Thinking, as a human-centred approach to design, has an appreciation for art:

> "Everything we have around us has been designed. Design ability is, in fact, one of the three fundamental dimensions of human intelligence. Design, science, and art form an 'and' not an 'or' relationship to create the incredible human cognitive ability." (Cross, 2006)
>
> "The most secure source of new ideas that have true competitive advantage, and hence, higher margins, is customers' unarticulated [i.e., emotional] needs." (Liedka, 2014)
>
> "The key to the process is empathizing with the users (citizens) to uncover unmet needs by understanding their beliefs, values, motivations, behaviors, pains, gains and challenges [i.e., their emotional needs] and to provide innovative solution concepts." (Hwa et al, 2020)

In a nutshell, one may reiterate that, "While the practice of engineering may involve as much technical experience as the poet brings to the blank page, the painter to the empty canvas, or the composer to the silent keyboard, the understanding and appreciation of the process and products of engineering are no less accessible than a poem, a painting, or a piece of music" (Petroski, 1992)—To engineer is as *human* as playing music or painting. The "human-centred" system design ought to combine technical and artistic "requisites". Thus, Pagani got it right with the design and development of the *Huayra*, an advanced car that seeks to satisfies the engineering *as well as* the artistic requirements of the customers.

12.2 The Hidden Costs of Design Mistakes

It has transpired from the author's recent interactions with practitioners across the LIPs industry that many of them have now embraced a definition of project success that goes beyond the traditional "Iron Triangle" (i.e., the tripartite constraints). Projects should *firstly* improve operations and, thus, derive benefits to the business, in addition to being satisfactorily completed on time, on budget, and to specification.

The "Iron Triangle" has yet another problem: it is not equilateral (contrary to what some PM literature may suggest). Cost, time, and specifications do not carry the same weight vis-à-vis project success. For instance, whereas budget overruns might not necessarily bring about schedule overruns, any errors or inadequacies in design specifications (i.e., engineering errors or inadvertences) will most likely entail both cost and time overruns.

Regarding project delays, one will probably identify one or two large projects (e.g., power plants, railways, pipelines, etc.) that are being overly delayed due to flawed design, poor development planning, construction incompetence, or a rush to execution (i.e., moving to site *before* design is complete). Though we often laugh about it for our own sanity (and perhaps we should), we ought to appreciate that megaprojects being delayed due to engineering mistakes (or inadequacies) might prove detrimental to the economy. Many developing nations were wrecked due to or following these kinds of project misfortunes.

In addition to technology and design failures leading to failed construction and/or chronic underperformance during subsequent operations, they cause delays to infrastructure megaprojects and impair their economics as follows:

(1) Increased in supervisory work and consequently in man-hours, which translates into more fees to consultants, and further payments to contractors and subcontractors
(2) Escalations and inflation costs, due to increases in prices of goods and/or services in the general economy, and particularly in the construction and related industries
(3) Delayed operational and business benefits and/or missed revenues, as a result of delayed operational improvements (e.g., provision of public/essential services)
(4) Opportunity costs due to other investments being delayed, interests forfeited or loss of goodwill, and more importantly, loss of potential clients/customers by the time the facility is eventually complete.

These financial repercussions are debilitating to the economics of projects, to the business, and to the country's overall economy. They might even destroy "value" in the projections of the Net Present Value (NPV), Income Statement, and Balance Sheet. The author provides a succinct discussion as to how "*project, business, and economy*" may be gravely affected by delays, whenever such prove material in quantum (i.e., magnitude) or in frequency.

I. ***How do engineering delays affect the project?***

Prolonged cash out-flows (e.g., increase in total outlay) and delayed cash in-flows (i.e., delayed income or revenue) adversely affect the NPV of the candidate project, turning its NPV into a lesser and/or negative amount and therefore inflicting a reduced or adverse financial viability to the project:

$$NPV = -Total\ Outlay + \sum_{j=year\ 1}^{j=project\ span} \frac{\left(Income_j - Expenditure_j\right)}{(1 + Discount)^j}$$

Prolonged cash out-flows (i.e., increase in total outlay) and delayed cash in-flows (i.e., delayed income or revenue) also deteriorate (i.e., increase) the Pay-Back Period (PBP) of the candidate project. For the longer the PBP, the riskier the investment, particularly in terms of its exposure to market and other operational risks:

$$PBP = J_{(years)}\ such\ as \left[0 \geq \left(-Total\ Outlay + \sum_{j=year\ 1}^{j=project\ span} Income_j\right)\right]$$

The deterioration of either NPV or PBP will translate in reduced financial viability of the project and a reluctance on the part of the project sponsor and/or potential investors (e.g., financiers, infrastructure development agencies, etc.) to invest in such a project, should alternative candidates be available (including simply walking away from the non-viable project).

II. ***How do engineering delays affect the business?***

Depending on loan payment clauses, raising monies to pay off the debt (i.e., annual instalments) as the facility remains under *work in progress* (due to delays and, thus, not generating revenue) adversely affect the Income Statement, worse still, when capital outlay is required to complete the project. This will spell bankruptcy (and fire-sale of the company's assets), should the business not be able to raise additional capital.

Similarly, delayed completion may expose the business to a *bleeding* or *collapsing* Balance Sheet by reflecting project-induced liabilities without a completed asset to show for it. In addition, this can and will often lead to a deterioration of the company's Debt Ratio, Return on Capital Employed (ROCE) and, finally, of its credit worthiness (hence, sharp increase in costs of borrowing). Moreover, according to International Accounting Standards (IAS, Section 23), any "unjustified" delays (e.g., not resulting from natural disasters, force majeure, acts of God, etc.) could cause the devaluing of the facility/asset due to the forfeiting of the capitalisation of borrowing costs.

These financial predicaments, if not adequately addressed, may well be exacerbated by operability and commercial failures, should the facility/asset fail to secure and/or service its customers by the time it is eventually complete. The Brazilian government and soccer federation abandoned a stadium project halfway through as soon as they realised it might not be ready for the 2014 Soccer World Cup. They were wise enough to cut their losses, but averting delays *in the first place* would have been much wiser.

III. ***How do engineering delays affect the economy?***

Projects are initiated and funded by government agencies and/or private companies. In either case, when those entities are adversely affected by the project delays (as discussed above), the cumulative effects on the economy can only be negative due to higher prices/tariffs on goods and services (i.e., in an attempt by most companies to reverse a "negative" bottom line on their Income Statement), sluggish GDP growth, job losses, forfeited taxes, increased imports, delayed exportations, and delayed operational improvements in infrastructure-supported services—A very bad scenario!

At a national level, especially when such delays apply to LIPs, the above economic predicaments affect the balance of payment, the sovereign credit rating, the cost of doing business, and the ability to attract foreign direct investments. Further, the expected improvements in the standard of living of citizens, the accumulation of national wealth and the country's overall competitiveness will suffer. Indeed, just one "delayed" project was enough to wreck both the economy and sovereignty of Egypt.

History suggests that modern Egypt nearly went bankrupt as a nation (circa 1875) and eventually fell under British rule, courtesy of one "challenged" project, the Suez Canal—the difficulty arose from its deficient design (i.e., "How to build"). It took 10 years to build the 165 km canal that joined the Mediterranean and Red Seas in 1869. As a result of delays, the final cost of the canal came in with an overrun of 167%. Most such delays stemmed from the engineering challenges pertaining to excavations, which contractors only addressed at the later part of the project (as of 1865) by funding further developments of a "steam ladder dredge" and, thus, adding to cost overruns.

Much closer to our own times, *The Washington Post* (USA) wrote on 25 April 1982: "Development in the Third World is never an easy task and is usually fraught with perils. Debt-ridden Zaire [now called DR Congo] is often regarded as a case study in Western-assisted development gone [terribly] wrong [...] the US government and American banks have helped finance a billion-dollar power line to carry electricity to an area that does not need [much of] it" (The Washington Post, 1982).

The article further notes, "Zaire has spent $1 billion to construct the 1,100-mile Inga-Shaba power line, which is scheduled to start operations later this year. It is the longest direct current line in the world [Well, it was so until 2010], and a Western diplomat described the technology as '22nd century'. Electricity is to be carried from generators at Inga near Kinshasa to the southern Shaba [also called Katanga] province in the heart of Zaire's copper belt, where the power will be converted to alternating current. The problem is that Shaba has no use for the [Inga] electricity, while power-starved communities along the line [e.g., in the Kasai provinces] are unable to tap the direct current without building expensive [and unreliable] converter stations" (The Washington Post, 1982).

The "*22nd century technology*" proved rather defective, with inverter designs based on "unproven" 12-Pulse-Thyristor valves. The line operated initially at only 10% of capacity [...] As of year 2000, reliability had been poor with many thyristor failures at the inverter station. Since electricity is not reaching the people living near the powerline, they tend to find their own uses for it, e.g., removing parts of the metal towers to make beds, shovels and other equipment. With an overwhelming cost overrun of about $500 million, the grandiose $1 billion project has so far turned out to be a white elephant (The Washington Post, 1982); thus, adding to the woes of a nation "already strapped with $4.5 billion of debt"—from wrong/flawed engineering designs to cost and schedule overruns to (further) economic woes; a "harmful" cycle that could have been avoided.

As if history will never cease to repeat itself in Africa, the world is now waiting to see what might happen to the economy (and sovereignty) of South Africa, where two of the world's largest coal-fired power plants (Medupi and Kusile) are being constructed. Both projects are 15 years behind schedule—again, due to engineering inadvertences. Meanwhile, the country is plagued with *crippling* power outages as well as *steep* tariff hikes.

"He [e.g., design engineer] who has ears to hear, let him hear." [Luke 8:8]

12.3 Practical exercise no. 1

Poor Systems Engineering has been blamed for system acquisition failures (i.e., schedule and cost overruns, cancellations, or delivered "systems" that are not fit for purpose – i.e., operations not improved). On 10th July 2020, Business Insider wrote about "design flaws" at Medupi, the fourth largest coal-fired plant in the world (Business Insider SA, 2020):

> *"There is [an] increased risk of load shedding [over] the next twelve months", André de Ruyter [new CEO of Eskom] said in an interview with The Money Show on 702 and Cape Talk on Thursday evening. He said that the company has to do urgent "reliability maintenance" [i.e., maintenance interventions aimed at increasing "mean time between failure" of parts] at its older power stations, which means they will have to be offline for long stretches of time [or] otherwise the risk of load shedding [i.e., power outages] will only increase in coming years.*
>
> *The average age of Eskom's coal-fired power stations (excluding Medupi and Kusile) is 38 years. The maintenance on these plants should be completed by the end of winter 2021. But that's not the only problem. On Tuesday, a unit at the brand-new Medupi tripped [i.e., for the nth time].*
>
> *"It is disappointing to see the performance of this new plant", de Ruyter says. Eskom has started to fix the design flaws at all twelve units of Medupi and Kusile (six each). It costs R 300 million [i.e., US$ 20+ million] to fix one unit, and also means that each unit is out of action for more than four*

months [thus, loss of revenue] [...] Fixing the design flaws involves "enormous" changes to each unit, including the removal of 32km of boiler tubing, says de Ruyter. The existing boilers [i.e., with inadequate height] malfunction because they burn too hot. He says the boilers did not take into account the properties of the coal that is being mined by Exxaro's Grootegeluk Mine for the plant.

De Ruyter said South Africa was in too much of a hurry when the new plants were commissioned. "We took an off-the-shelf design and are now paying the [ultimate] price in terms of a plant not suited to coal that is available for it to burn".

With US$ 30 billion in liabilities, Eskom is the biggest risk to the South African economy; was the new CEO uttering his *epiphany* (i.e., realisation of the task ahead), or giving an *epitaph* (i.e., assent of death) about Medupi?

Question 1:

- Is Medupi a "failure" of project management, of engineering, or both?
- Discuss its commercial viability, given the ramp-up delays and cost overruns.
- Discuss what other "systems" a possible failure at Medupi might also affect?

Question 2:

Each boiler being a "system", discuss the proposed solution to design flaws:

- How would you explain that such design flaws were not detected sooner?
- Would the replacement of boiler tubing possibly create more problems?
- What organisational functions ought to improve to avert another fiasco?

12.4 Practical exercise no. 2

You are required to discuss the account of the steamboat *Sultana* catastrophe (1865) based entirely on the information provided, as summarised in the two excerpts below:

(a) *"On the night of April 26, 1865, the passenger steamboat Sultana exploded on the Mississippi River seven miles (11 km) north of Memphis, Tennessee. This maritime disaster is categorised as the worst in the history of the United States. The explosion resulted in the tragic loss of 1,547 lives, surpassing the total number of deaths caused by the sinking of the Titanic (i.e., 1,503 fatalities) [...] The Sultana was overcrowded due to a soldier-prisoner exchange [...] The explosion happened around midnight which was when the Mississippi River was at flood stage. It was documented that the single metal lifeboat on board the Sultana was thrown from the upper deck, landing on several people swimming away from the steamer, resulting in further deaths."* (Berryman et al, 1988)

(b) *"The disaster was believed to be the result of a repaired boiler explosion that led to the explosion of two of the three other boilers. The initial boiler had been previously found to have had a leak and was improperly repaired by boilermaker R. G. Taylor due to orders from Captain J. Cass Mason because of time constraints in Vicksburg, Mississippi. While chief engineer Nathan Wintringer approved the repaired boiler, Taylor stated that the boiler could not be considered safe for operation since the boiler appeared to be burned from being worked on with too little water."* (Berryman et al, 1988)

A boiler is a fuel-burning apparatus or container for water heating or for central heating system. **It is also used to produce steam for power generation**. The generation part of a steam system uses a boiler to add energy to a feedwater supply to generate steam. The "heat energy" is released from the combustion of fossil fuels or from process waste heat.

Question 1:

From the Engineering Design and Development perspective and/or process:

- Who do you suggest should have been held responsible for this disaster?
- What could you have done to prevent it, had you been part of *R. G. Taylor*?
- What lessons have you learned from this unfortunate experience?

Question 2:

Assuming you were in charge (e.g., manager) of infrastructure delivery in your organisation:

- What organisational functions and/or structures will you put in place to prevent such detrimental occurrences from happening in the first place?
- Where should your focus lie: on the Planning, Design, Construction or Operations?

12.5 Practical exercise no. 3

The Cambridge Dictionary (1995) defines the term ecology as, "*The relationship of living things to their environment and to each other, or the scientific study of this*". Jamshidi (2009) similarly remarks that, "The network, its enabling services, and the systems connected to it create a network ecology or business/operational environment that enables new types of systems to become viable, a form of 'emergent properties' thought to characterise an SoS".

Question 1:

Based on the above definitions, consider a particular ecology consisting of a number of transportation systems (i.e., road and/or railway network) and their PESTEL (Political-Economic-Social-Technological Environmental-Legal) Context:

- The sea port as a "system" is made up of many terminals (e.g., container, passenger, oil/gas, and coal terminals) that are independently managed
- The Design Process shall consist of the following steps/phases:
  i. Consider the envisaged port as an "interface" between two or more ecologies still to be identified by the project team based on the general geography
  ii. Consider how to pertinently enhance and/or expand the ecologies and interfaces involved (i.e., what and how to transport). This exercise could spawn a few infrastructure projects and/or PESTEL interventions
  iii. Consider the insights from the previous steps to determine a suitable "port configuration" (e.g., what specific terminals to cater for) and capacity (i.e., size of berths, operability) required to achieve a set of objectives.

- The team shall assume operational objectives that are acceptable to management.

Question 2:

Use this process to outline the "upgrade" of Maputo International Airport.

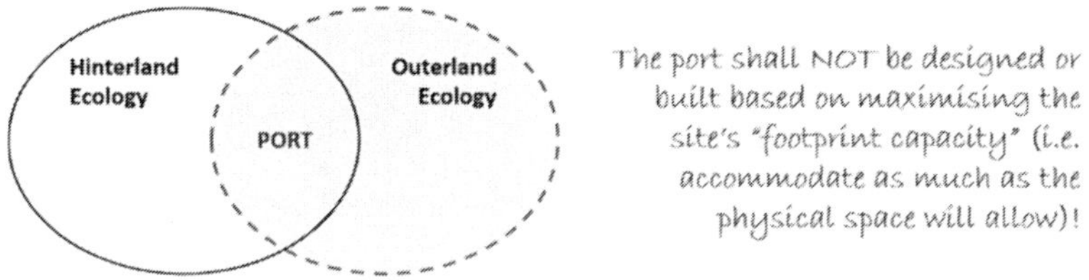

Figure 12.1 – Example of a Generic Port Ecology

12.6 Practical exercise no. 4

Robben Island is quite a notorious site in South Africa; it has been "home" (i.e., lifetime jail) to scores of political prisoners during Apartheid. It is now being turned into a top tourism destination for both locals and foreign visitors. A permanent crossing to Cape Town, the *Cape-to-Robben-Island* Link (6.9 km), has been proposed to reduce the time and costs, and improve the safety of the trip. Discuss likely client and system requirements and their Functional Allocation.

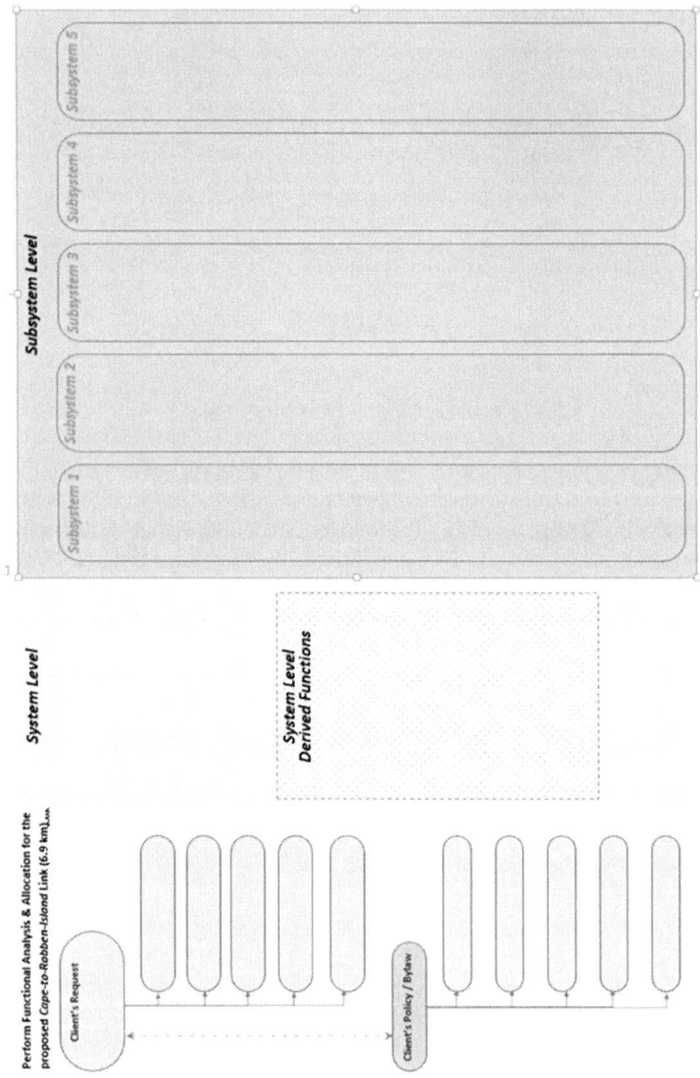

Figure 12.2 – Functional Analysis and Allocation Template

12.7 Practical exercise no. 5

Your team has been appointed to perform a Functional Analysis and Allocation exercise for the anticipated 6.9 km *Cape-to-Robben-Island* "permanent link" (see Figure 12.2).

Question 1:

Considering the System/Facility (as previously analysed), please discuss and develop the following: (1) Functional BS, (2) System BS, (3) Contracting Line, and (4) Roles and Skills expected of the Owner's Team in the template below (see Figure 12.3).

Question 2:

- What insights would you think developing a Work BS could have added?
- What could still be learned from the "relationships of the breakdowns"?
- How would you reconcile those "relationships of breakdowns" to the three sections of the Systems Engineering Management Plan or SEMP?

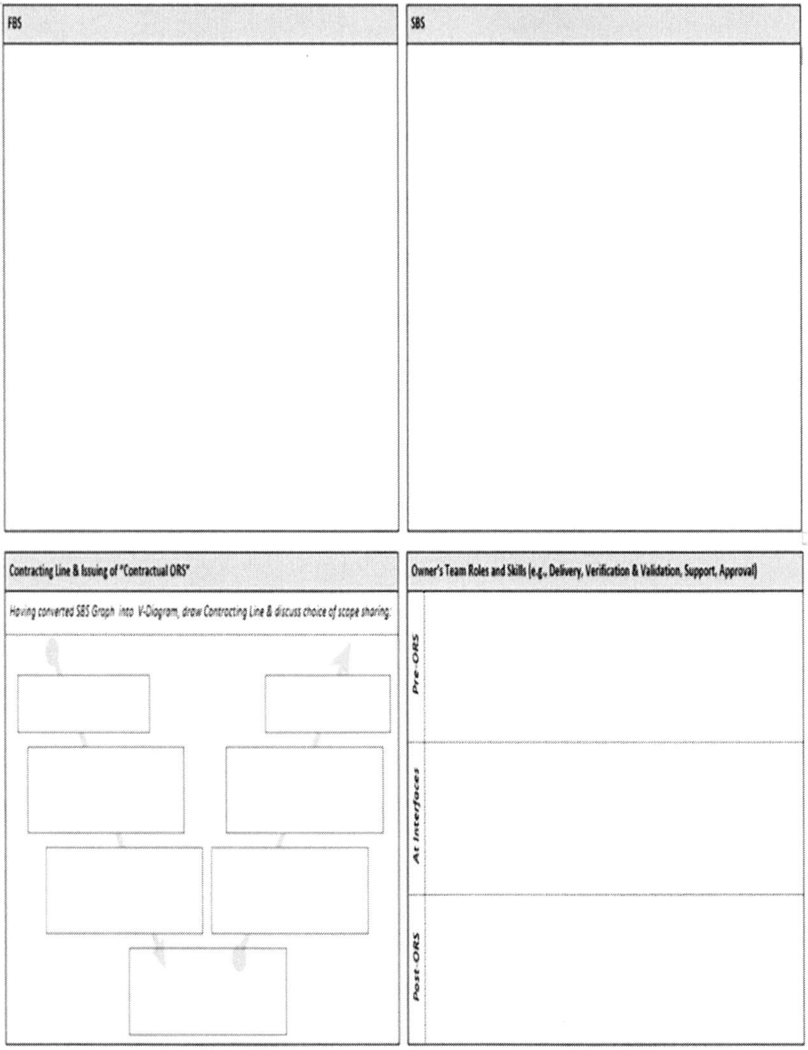

Figure 12.3 – Template for FBS, SBS, Contracting Line, and Roles and Skills

12.8 Practical exercise no. 6

Considering the SBS for the proposed *Cape-to-Robben-Island* Link [Practical exercise no. 4 above], discuss and develop the WBS using the WBS Template provided here.

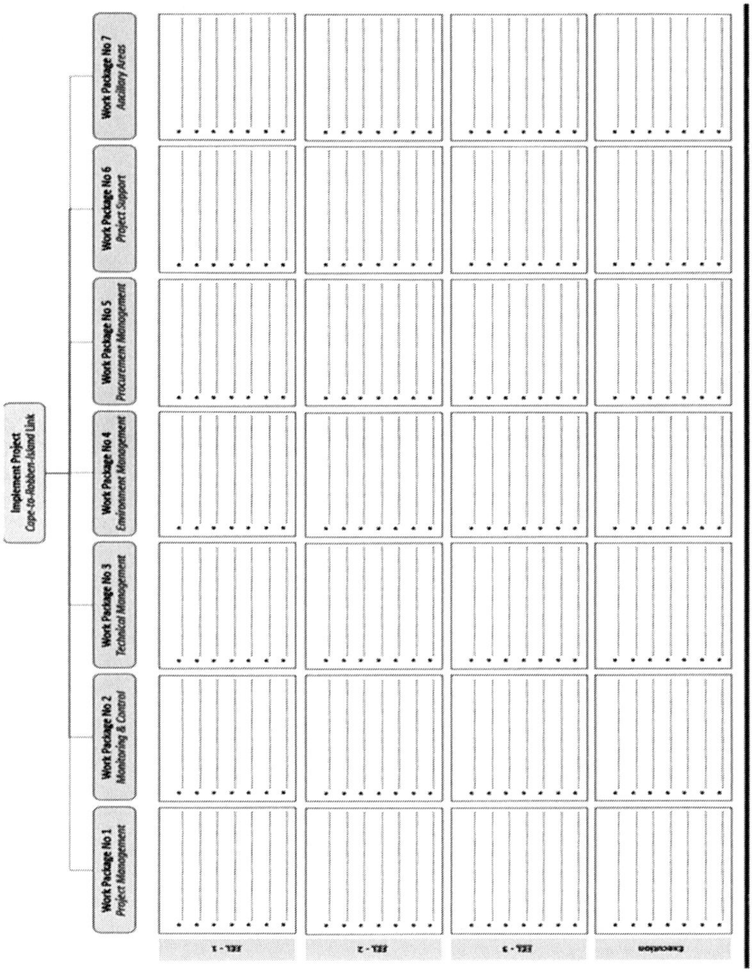

Figure 12.4 – Work Breakdown Structure (WBS) Template

12.9 Practical exercise no. 7

"Engineering Managers have to be involved in the management of product [i.e., system] design and development either as part of a designing team or as a purchaser of a [large infrastructure] system that is being designed." (Sivaloganathan and Al-Marzouqi, 2014)

You are leading the engineering team for the *Cape-to-Robben-Island* project; as part of the job, you are expected to consider the approaches to Engineering Management below:

(i) "Systems Engineering Management is a technical function and discipline that ensures that systems engineering and all other technical functions are properly applied [in the design]"—NASA SE Handbook (Systems Engineering Handbook, 2007).

(ii) "Larger [and complex] projects have historically experienced high levels of rework with accompanying cost and schedule impacts because design and construction elements have been allowed to proceed in advance of full requirements definition and/or without adequate information on site conditions, operating environments, market capabilities, etc."—Managing Design and Construction Using Systems Engineering, US Department of Energy (Roy, 1993).

(iii) "Unless we look at things as systems that are somewhat interconnected, we wouldn't perceive anything beyond their faint and fleeting shadows" (The author).

Question 1:

Discuss these Systems Engineering approaches in the context of Design and Development (EDD) for the aforementioned *Cape-to-Robben-Island* "Permanent Link" project.

Question 2:

The SEMP (Systems Engineering Management Plan) outline is to be tailored to the development of a system and may include the following elements/aspects:

I. **Introduction**
   - Scope
   - Purpose
   - Overview
   - Applicable Document

II. **Programme Planning and Control:**
- Organisational Structure
- Responsibilities
- Procedures
- Authorities
- WBS
- Milestones, Schedules, Programme Events, Programme
- Technical
- Test Readiness Reviews
- Technical and Schedule Performance Metrics
- Engineering Programme Integration
- Interface Plans

III. **Systems Engineering Process:**
- Mission
- System Overview Graphic
- Requirements & Functional Analysis, Trade Studies, Analysis of Alternatives
- Technical Interface Analysis/Planning
- Specification Tree/Specifications
- Modelling and Simulation
- Test Planning
- Logistic Support Analysis
- Systems Engineering Tools

IV. **Engineering Integration:**
- Integration Design/Plans
- Compatibility/Interference Analysis
- Specialty Engineering (i.e., – *ilities* Engineering)
- Constructability/Producibility Studies

Applying "generic" SEMP provided above, answer the following questions:

(a) Building on the previous exercise [*Cape-to-Robben-Island* Link project], discuss the essence of the applicable elements/sections of the SEMP
(b) Considering the difference in purpose between FEL-2 and FEL-3, discuss how such differences could manifest themselves in or around the SEMP

Question 3:

- Assuming that the SEMP is effectively developed, discuss how you might apply a Monitoring and Control tool such as the Earned Value Management System (EVMS) or any similar tools to assure its successful implementation
- Discuss how often you would expect to update the SEMP during Construction

Question 4:

How does the SEMP align to a "Delivery Plan" containing the following sections?

(A) **Context and Purposes:**
The context, importance, rationale (Why), and objectives/goals of the project.
Two other spheres/levels ought to be considered – one level higher and lower.

(B) **Scope:**
The areas of focus, activities and deliverables expected from the project.
The totality of the scope and any scope by others and/or owner listed as such.

(C) **Timelines:**
The sequencing of and dependencies between major activities.
The key milestones (e.g., reviews, inspections, and/or approvals points).

(D) **Human and Other Resources:**
The skills, capacity and organisational structures (e.g., roles, teams, interfaces).
Any special equipment, (satellite) office space, travel, special insurance covers.

(E) **Budget:**
The breakdown of budget and/or cashflow, as well as any "interim" income.
The sources of funding, cost of borrowing, escalations, FOREX components.

(F) **Major Risks:**
The high-level threats and opportunities arising from the project.
The implications to organisation (e.g., policy, legal, reputation) or environment.

12.10 Practical exercise no. 8

By defining the candidate system to be designed in terms of its hierarchy (i.e., system-subsystem-assembly-element) the design will gain a better and deeper understanding of the system performance. This would provide two benefits:

i. Understanding how the "component's solutions" contribute to the designer's aptitude to detect underperforming components and/or interfaces and, therefore, the ability to draw upon "innovation" in order to improve such features.
ii. Understanding how the "component's problems" contribute to the designer's aptitude to troubleshoot poor performance or undesirable behaviour at the overall system level owing to their deep/enhanced knowledge of its broader architecture, i.e., its constituent elements and effectual interfaces.

Question 1:

Recognising the axiom: "*one person's system is another person's subsystem*", you are required to select any of the subsystems constituting a shopping mall (i.e., parking lot, shopping floors, aisles/passages, restrooms, facilities services). You shall proceed to develop a "Hierarchy of Systems" up to element-level; then you will discuss the *problems* and matching *solutions* as per the template. (*Do not hesitate to take a hint from the illustration provided in Table 3.3*).

Hierarchy	Problem(s)	Solution(s)	Component
1 System			
2 Subsystem			
3 Assembly			
4 Element			

Table 12.2 – System Hierarchy and Technical Solution Template

Question 2:

- Discuss what deeper or better appreciation of this facility you have gained.
- What are the implications of failing to define assembly-level "problems"?
- Having completed this exercise, should you hear that "*passages are being converted into open restaurants*" in that mall, what shall you advise them?

12.11 Practical exercise no. 9

A team of engineers is working on a railway project. The idea is to develop the fastest train in Africa that might hopefully compete with the best in the world. The team leader suggests applying the QFD method to design the "African Spear-Train" (i.e., AST 2040) and to plan the whole engineering development process. At this stage, investigations have revealed that the train will have to compete with an existing highway and an outdated 80 km/h trans-urban train line operating in the surrounding regions.

In accordance with the QFD method, the team undertook to identify customer needs pertaining to the "African Spear-Train" and their comparative importance. They also collated information about customer perceptions of competing "products" before proposing technical requirements. Further, the engineers determined the applicable technical correlations as well as their interrelationship with the customer requirements. The HoQ for Planning (Figure 12.5 below) captures all this information in a summary form to inform the design process.

Question 1:

Since applying the QFD method entails "giving a voice to the customer", discuss the plausible reasons (e.g., benefits to project, customers) that might have caused the project manager to elect the QFD.

Question 2:

Considering the introductory section on the QFD method (see Chapter 05) and the below "HoQ for Planning", develop and discuss the other HoQs, namely, the "HoQ for Parts", "HoQ for Process", and "HoQ for Production Planning".

Note: In answering these questions, please use the following "rule of thumb":

(i) Indexes for the Interrelationship Matrix (i.e., between Customer Requirements and Performance Measures) should range from 1 to 3 (weak), from 4 to 6 (moderate), and from 7 to 9 (strong) to denote the relationship's strength.
(ii) Negative impacts represent constraints, i.e., improving one of them may actually cause a negative impact on the other. Sometimes an identified change impairs so many other items that it is just better to leave it alone lest the system eventually exhibits a "Shifting of the burden" architype.

House of Quality

Quality Function Deployment for "Spear-Train Design"

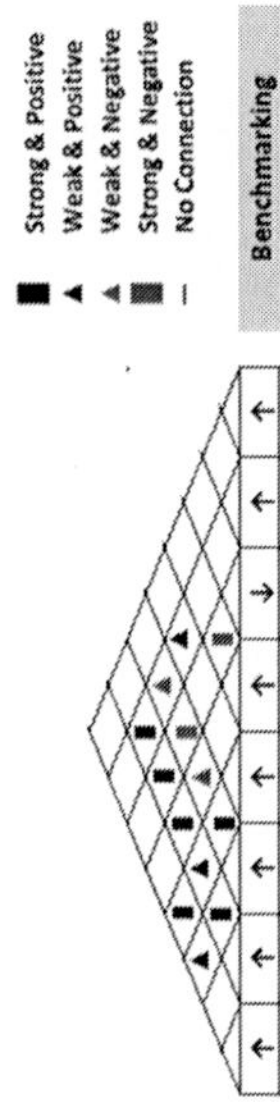

		Importance (Scale of 1 to 5)	Relative Importance (in %)	Relative Ranking	Width of Carriages ↑	Number of Traction Motors ↑	Maximum Axle Load ↑	Capacity of Carriages ↑	Turning Radius ↑	Maximum Rail Gradient ↓	Number of Luggage Wagons ↑	Number of Toilets per Wagon ↑	Existing Highway	Old 60 km/h Railway
01	Smoothness of Ride	3	13.6%	04		8	8	7	8	7			4	2
02	Speed of Travel	5	22.7%	01	2	9	6	9	7	8	4		5	1
03	Commuter Comfort	2	09.1%	06	8	6	4	4	5	6		6	3	1
04	Commuter Safety	4	18.2%	02	7		7	8	9	6			1	2
05	Inexpensive Trip	4	18.2%	02		7	8	6					1	3
06	"Unrestricted" Luggages	3	13.6%	04	3	5	6				9		1	3
07	Availability of Toilets	1	04.5%	07	2			3				9	2	1
	Technical Difficulty Involved				4	3	5	3	4	4	2	1		
	Measurements Unit				m	no off	t/axle	ton	km	degree	no off	no off		
	Object Target Value				2.2	2	20	25	12	3:1	3	2		
	Absolute Importance of Design Parameter				03.0	05.6	06.4	06.0	04.8	04.4	02.1	01.0	17	13
	Relative Importance of Design Parameter				6	3	1	2	4	5	7	8		

Benchmarking

Figure 12.5 – Example of House of Quality for Planning (HoQ$_1$)

12.12 Practical exercise no. 10

The NASA Technology Readiness Levels (TRL) suggest that any technology shall evolve to *maturity*, from the TRL 1 (basic concept) right up to the TRL 7 (proven technology), for an "unproven" technology would impair the engineering and economics of the project:

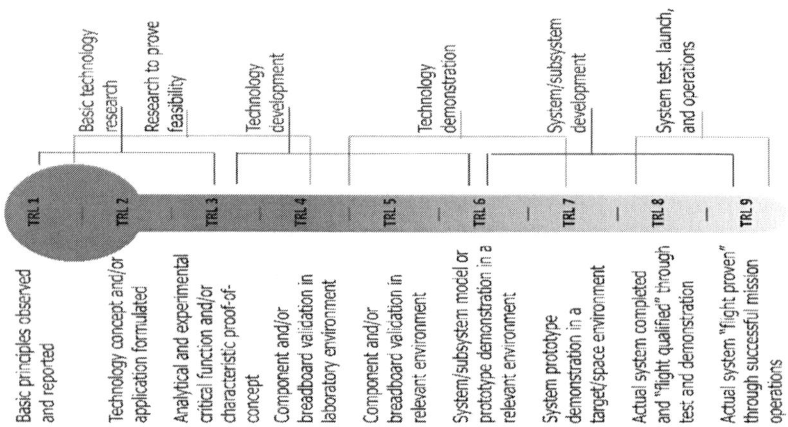

Figure 12.6 – Technology Readiness Levels (TRL)
(as per NASA SE Handbook, 2007)

Using the NASA TRL Scale, how would you assess the "Technology Readiness" of each of the "Space Craft XM-19" subsystems (i.e., module), and the System as a whole?

Subsystem		Component	#	01	02	03	04	05	06	07	08	09	10	11	12	13	14	15
Space Craft XM-19 Architecture		Component's TRL ↘																
001	Cruise and Launch	*Delta II Launch Vehicle*	01	9	X		X											
		Cruise Stage Structure	02	X	9	X	X											
		Cruise Power System	03		X	9							X					
002	Entry, Descent and Landing	*Aeroshell (Heatshield & Backplane)*	04	X	X		5	X			X							
		Parachute	05				X	6	X		X							
		Solid Rocket Assisted Decelerator	06					X	6		X							
		Airbags	07							6	X							
		Backshell Pyro Switching & Pyros	08				X	X	X	X	8	X						
003	Power Distribution	*Lander Shunt Limiter Unit*	09								X	9		X	X	X		
		Power Distribution and Control	10			X							9				X	X
		Lander Petals / Solar Array	11									X		9				
		Lander Battery	12									X			9			
		Lander Pyro Switching & Pyros	13									X				8		
		Cruise Power Converter	14										X				9	
		AIM Power Converter	15										X					9

Table 12.3 – Example of TRL for Space Craft XM-19 System

Question 1:

Which of the following SE Rules would you recommend in assessing the "Technology Readiness" of these Subsystems and System?

i. The TRL of the "weakest link" (i.e., element carrying the most immature technology) reflects the TRL of the System or subsystem under analysis
ii. The lowest TRL of "interfaced elements" (e.g., Subsystems, assemblies, components) reflects the TRL of the System/subsystem under analysis
iii. The TRL should be computed as per point (ii), but discounted to allow for the "novelty of combination" to reflect the TRL of the Subsystem.

Question 2:

The "Cruise and Launch" element may have reached TRL 9 as a standalone; would you justify a lower TRL being assigned as part of Space Craft XM-19?

Question 3:

Should the "Overall TRL" of a component and/or subsystem fall below the NASA recommended level, what would you as the senior project manager recommend as a *way forward* to the project's engineering manager and to the executive management? What might be the consequences otherwise?

Question 4:

Assuming this fictitious NASA Project was at Pre-Feasibility (FEL-2) phase, what kind of conclusion(s) would you draw as to its "Technical Feasibility"?

Question 5:

Assuming you are a "designer" by profession (not an astronaut), what lessons would you think you have or should have learned from this NASA exercise? Discuss what benefits you might have derived from applying these learnings to other infrastructure projects, especially to those large and complex initiatives?

Question 6:

- In terms of EDD life cycle, at what phase will you locate this "process"?
- Discuss how you would have guided the technical team on your project to apply this TRL tool on the shopping mall example [refer to Exercise no. 8] and discuss what design opportunities and challenges might have transpired.

12.13 Practical exercise no. 11

It is generally acknowledged that Design Thinking (DT) is a "human-centred" problem-solving approach that supplements other design methods, including QFD, with two crucial inputs as follows:

(1) The Empathise Phase provides the elicitation of "unmet, emotional and/or unarticulated needs" that are indispensable to creating the desired outcomes (as per the ConOps) that benefit users/customers.
(2) The Ideate Phase is all about unlocking potential and creating confidence in the devising of feasible and viable big ideas and solutions. It accommodates advances in innovation and efficiency because it is in the generation of "innovative" ideas that novel solutions are found.

Question 1:

Based on the above considerations, you are assigned to a team and expected to select and enhance a "product" and/or "service" of your choice as follows:

i. List any positive or negative characteristics of the product/service.
ii. Why would you deem any such characteristic as positive/negative?
iii. Describe the users and your own feelings about each characteristic.
iv. Based on points (i) to (iii), specify the product/service requirements.
v. Describe the proposed design (by way of sketch, prototype or model).
vi. Discuss the validation (i.e., desirability, feasibility, viability) of the design.
vii. How will you describe this experience as compared to a traditional process?

Question 2:

The head of infrastructure development of the Afro-diasporic "Kwanza State" has appointed your consulting firm to advise them on the requirements for the Road/Rail Ring (R3 Link) that will connect the federal capital to adjacent states:

- How shall your company apply DT in this particular situation?
- Will it perhaps be a good idea to apply DT to a sewer plant?
- What benefits will you sell to the client regarding applying DT?
- Who would you recommend should be involved in this "process" and why?
- Would you advise that a "facilitator" be appointed to lead this process?

12.14 Practical exercise no. 12

Having understood the "Four Levels of Requirements" according to Scott (2014) as discussed and applied to the engineering of a new *wristwatch* as a system in Chapter 03, you are part of the engineering team appointed to deliver a Juvenile Prison, "*The Rejuvel*". The team will have to discuss the system requirements, behaviours, architecture, and measurements pertaining to the envisaged Juvenile Prison, "*The Rejuvel*" (as a system).

Question 1:

Discussions on Systems Requirements should start by defining the "System Purpose":

- Is prison for locking away juvenile convicts, keeping them out of civil community?
- Is prison for punishing juvenile convicts, making them suffer for their misdeeds?
- Is prison for rehabilitating juvenile convicts, converting them into decent citizens?
- Is prison intended to fulfil a judicious combination of these aforesaid purposes?

Question 2:

What interactions (i.e., within the prison-system, with the outside world) will produce the "purposed" outcomes?

Question 3:

- What "components" are required (by design) in a prison to produce the "purposed" outcomes?
- What interfaces are required (or not) among components to create those outcomes?

Question 4:

How do you ascertain that the right Juvenile Prison is being built and/or operated correctly? Your team shall consider hard/soft value criteria provided in Table 12.4 and Table 12.5.

Question 5:

Considering the "Juvenile Prison" system as the epicentre of a broader system or SoS; what other systems or SoS (e.g., physical, social, economic) might be connected to it?

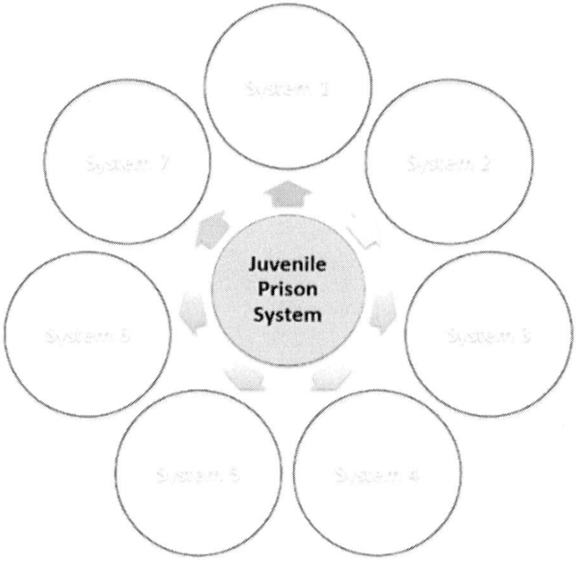

Figure 12.7 – Systems Interconnecting to the Juvenile Prison System

Examples of Primary Value [Hard Data]			
Cost Savings	***Time Savings***	***Increased Output***	***Quality Increases***
Accident Costs	Administrative Time	% of Traffic Increase	% Tasks Completed Correctly
Administrative Costs	Average Delay Time	Applications Processed	Customer Complaints
Average Cost Reduction	Complaint Response Time	Forms Processed	Deviation From Standard
Budget Variances	Cycle Time	Incentive Bonus	Error rates
Cost By Account	Efficiency (i.e., Time)	Inventory Turnover	Inventory Adjustments
Delay Costs	Equipment Downtime	Items Assembled	New Products Created
Fixed Costs	Late Reporting	Items Sold	No of Accidents
Operating Costs	Lost Time Days	Loans/Bids Approved	No/Cost of Complaints
Overhead Costs	Meeting Time	New Account Opened	No/Cost of Returns
Overtime Costs	Order Response Time	Patients Visited, Treated or Discharged	Processes Upgraded or Reengineered
Penalties/Fines/Bonuses	Overtime	Productivity	Product Defects
Programme Costs	Processing Time	Shipments Completed	Product Failures
Project Costs Savings	Repair Time	Students Graduated	Rejects
Sales Expenses	Supervisory Time	Tasks Completed	Rework
Unit Costs	Time to Project Completion	Tons Manufactured	Scrap Created
Variable Costs	Training Time	Units Produced	Shortages
Write-offs/Bad Debts	Work Stoppages	Work Backlog	Waste

Table 12.4 – Examples of Primary Hard Values (applicable to various industries)

Examples of Primary Value [Soft Data]				
Initiative / Innovation	***Customer Service***	***Work Habits***	***Employee Development***	***Work Climate /Job Satisfaction***
% of Successful Projects	Customer Loyalty	Absenteeism or Tardiness	Increases in Job Effectiveness	Average Tenure or Staff Turnover
% of Suggestions Implemented	Customer Satisfaction Index	Excessive and Long Breaks	Inter-departmental Transfers	Discrimination Charges
Implementation of New Technologies	Customer Spread/Variety	First-Aid (On-Site) Treatments	No of Pay/ Salary Increases	Grievances and Litigations
Lessons Learned Logged/ Published	Customer Value Proposition	Insubordination and Theft/Fraud	No of Promotions Effected	Job Satisfaction Index
New Patents and Copyrights	Customers Lost versus Gained	Communi-cation Breakdowns	No of Trainings Attended	Multi-disciplinary Teamwork Index
New Products or Services launched	No of Customer Complaints	No of Safety Incidents	Performance Appraisal Ratings	No of Industrial Actions
Setting of New Goals/ Objectives	Repeat Business Deals/ Contracts	Visits (Repeated) to Dispensary	Request for Study /Sabbatical Leave	No of Repeat Hire (Re-Employment)

Table 12.5 – Examples of Primary Soft Values (applicable to various industries)

PERSONAL REFLECTIONS

BIBLIOGRAPHY

1 Ackoff, R., 2004. Transforming the Systems Movement. Presented at the third international conference on Systems Thinking in Management, 26 May 2004. Available at https: <//thesystemsthinker.com/wp-content/uploads/pdfs/150801pk.pdf>, [Accessed 24 May 2020].

2 Al-Mazrooie, J. and Bajracharya, A., 2013. A Study on the Operational Readiness of a Mega Construction Project. In: Creative Construction Conference. [online] Available at: <https://www.researchgate.net/publication/268332962_A_Study_on_the_Operational_Readiness_of_Mega_Construction_Project> [Accessed 30 May 2020].

3 Araghi, S., 2016. Lean Manufacturing Definition and Objectives. [online] Linkedin.com. Available at: <https://www.linkedin.com/pulse/lean-manufacturing-definition-objectives-sina-namaki-araghi/> [Accessed 30 May 2020].

4 Archibald, R., Di Filippo, I. and Di Filippo, D., 2012. The Six-Phase Comprehensive Project Life Cycle Model Including the Project Incubation/Feasibility Phase and The Post-Project Evaluation Phase. [online] Pmworldlibrary.net. Available at: <https://pmworldlibrary.net/wp-content/uploads/2013/01/PMWJ5-Dec2012-ARCHIBALD-DI-FILIPPO-Featured-Paper1.pdf> [Accessed 30 May 2020].

5 Australian Government, 2012. Infrastructure Planning and Delivery. 2nd ed. Canberra: Dept. of Infrastructure and Transport.

6 Bahill, T. and Madni, A., 2017. Trade-off Decisions in System Design. 1st ed. Switzerland: Springer, Cham, p.476.

7 Bar-Yam, Y., 2014. When systems engineering fails – toward complex systems engineering. International Journal of System Engineering. Cybernet. 2. 2021 - 2028 vol.2. 10.1109/ICSMC.2003.1244709.

8 Bellingham, R., 2001. The Manager's Pocket Guide to Corporate Culture Change. Amherst, Mass.: HRD Press.

9 Berryman, H., Potter, J. and Oliver, S., 1988. The Ill-Fated Passenger Steamer Sultana: An Inland Maritime Mass Disaster of Unparalleled Magnitude. Journal of Forensic Sciences, 33(3), p.12500J.

10 Bersson, T., Mazzuchi, T. and Sarkani, S., 2012. A FRAMEWORK FOR APPLICATION OF SYSTEM ENGINEERING PROCESS MODELS TO SUSTAINABLE DESIGN OF HIGH-PERFORMANCE BUILDINGS. Journal of Green Building, 7(3), pp.171-192.

11 Billah, K.; R. Scanlan, 1991. "Resonance, Tacoma Narrows Bridge Failure, and Undergraduate Physics Textbooks" (PDF). American Journal of Physics. 59 (2): 118-124. Bibcode:1991AmJPh.59.118B. doi:10.1119/1.16590.

12 Biswas P., Example of Process Failure Mode and Effect Analysis, APB consultant, available at <http://isoconsultantpune.com/wp-content/uploads/2015/01/Example-of-Process-FailureMode-and-Effect-Analysis.pdf.> [Accessed 05 Sep 2020]

13 BKCASE Editorial Board, 2011. The Guide to the Systems Engineering Body of Knowledge (SEBoK), v. 0.5. R.D. Adcock (EIC). Hoboken, NJ: The References Page 2 of 10, Trustees of the Stevens Institute of Technology. [Accessed 28 June 2016] www.sebokwiki.org.

14 Borza, J., 2011. FAST Diagrams: The Foundation for Creating Effective Function Models. In: Trizon Detroit 2011. [online] Available at: <https://www.aitriz.org/ documents/TRIZCON/Proceedings/2011-06_FAST-Diagrams-The-Foundation-for-Creating-Effective-Function-Models.pdf> [Accessed 30 May 2020].

15 Brown, T., Blog of 7 September 2008. Available at <https://designthinking.ideo.com /blog/definitions-of-design-thinking> [Accessed 14 February 2021]

16 Business Insider, 2020. Eskom CEO: Higher Risk of Load Shedding In 2020 – As 32Km of Boiler Tubing Removed From Medupi Unit. [online] Available at: <https://www.businessinsider.co.za/eskom-higher-risk-of-load-shedding-2020-7> [Accessed 15 August 2020].

17 Cambridge Dictionary, 1995. Cambridge International Dictionary of English, Cambridge University Press.

18 Canton Tower, 2010. “The Canton Tower”. Available at <http://www.gztvtower.info>. Retrieved 16 August 2020.

19 Carlson, C., 2012. Understanding the Fundamental Definitions and Concepts of FMEAs. Effective FMEAs, pp.21-55.

20 Chan, S., 2001. Complex Adaptive Systems. In: ESD.83 Research Seminar in Engineering Systems. [online] Massachusetts. Available at: <http://web.mit.edu/esd.83/www/notebook/Complex%20Adaptive%20Systems.pdf> [Accessed 30 May 2020].

21 Complex Systems thinking explained uncomplicatedly (video from www.healthytogether.vic.gov.au).

22 Cross, N., 2006. Designerly Ways of Knowing. London: Springer.

23 Crouhy, M., Mark, R. and Galai, D., 2009. *The essentials of risk management.*

New York, NY: McGraw-Hill.

24 Da Silva, L. and E Abreu, O., 2010. An IT infrastructure patterns approach to improve IT service management quality. In: *2010 Seventh International Conference on the Quality of Information and Communications Technology*. Irsee: IEEE, pp.171-176.

25 Dam, R. and Yu Siang, T., 2020. 5 Stages In The Design Thinking Process. [online] The Interaction Design Foundation. Available at: <https://www.interaction-design.org/literature/article/5-stages-in-the-design-thinking-process> [Accessed 15 June 2020].

26 Daylight, 2014. What Is Design Thinking. [video] Available at: <https://www.youtube.com/watch?v=Ee4CKIPkIik> [Accessed 4 July 2020].

27 Deutsche Welle, 2013. Flight Cancelled - The Berlin Airport Fiasco. [video] Available at: <https://www.dw.com/en/flight-cancelled-the-berlin-airport-fiasco/av-16512853> [Accessed 16 August 2020].

28 Earley, J., 2016. The lean book of lean – a concise guide to lean management for life and business. 1st ed. West Sussex, Wiley.

29 Edelman, G., 1987. Neural Darwinism. 1st ed. New York: Basic Books.

30 Eduardo, C., Dergint, D. and Hatakeyama, K., 2000. Project-based organizations as complex adaptive systems. 2004, pp.1-12.

31 Ferguson, A., 2014. MSP for dummies. Chichester, England: John Wiley & Sons.

32 Forsberg, K., Mooz, H. and Cotterman, H., 2005. Visualizing project management. 3rd ed. New York: John Wiley & Sons.

33 Fossnes, T. and Forsberg, K., 2006. Systems Engineering Handbook: A Guide for System Life Cycle Processes and Activities. 3rd Ed. [ebook] Seattle: INCOSE. Available at: http://disi.unal.edu.co/dacursci/sistemasycomputacion/docs/SystemsEng/SEHandbookv3_2006.pdf [Accessed 29 Apr. 2018].

34 Galvez, R., Gaylor, S., Young, C., Patrick, N., Johnson, D. and Ruiz, J., n.d. The Space Shuttle And Its Operations. [Online]. Available: <https://www.nasa.gov/centers/johnson/pdf/584722main_Wings-ch3a-pgs53-73.pdf> [Accessed 30 May 2020].

35 Glass, R. J., Ames, A. L., Stubblefield, W. A., Conrad, S. H., Maffitt, S. L., et al., 2008. Sandia National Laboratories: A Roadmap for the Complex Adaptive Systems of Systems (CASoS) Engineering Initiative, Sandia National Laboratories SAND 2008-4651.

36 Gottinger, H.W. 2012, Coping with Complexity: Perspectives for Economics, Management and Social Sciences, Springer, Netherlands.

37 Gould, S. J, 1996. The Mismeasure of Man. 1st Ed. US: Norton.

38 Habayeb, A. R., 2005. Architecture-Based Systems Engineering and Integration, 8th Annual Systems Engineering Conference Oct. 24-27, 2005, San Diego, CA.

39 Haik, Y. and Shahin, T., 2011. Engineering Design Process. Stamford, CT: Cengage Learning.

40 Hamil, D., 2002. Your Mission, Should You Choose To Accept It: Project Management Excellence. [online] Proceedings.esri.com. Available at: <https://proceedings.esri.com/library/userconf/proc01/professional/papers/pap529/p529.htm> [Accessed 15 June 2020].

41 Haskins, C. and Forsberg, K., 2011. Systems engineering handbook. Seattle, WA: INCOSE.

42 Hauser, J. and Clausing, D., 1988. The House of Quality, Harvard Business Review, Vol. 66 No 3.

43 Hellriegel, D. and Slocum, J., 1996. Management. Cincinnati, Ohio: South-Western College Publishing.

44 High Resolution, 2017. #16: IDEO's Tom Kelley Is Design Thinking's Ultimate Disciple, He Makes The Case As To Why. [video] Available at: <https://www.youtube.com/watch?v=L1pBhHjGKvI> [Accessed 4 July 2020].

45 Hitchins, D. 2009. "What Are the General Principles Applicable to Systems?" INCOSE Insight, 12(4): 59-63.

46 Honour, E., 2004. 6.2.3 Understanding the Value of Systems Engineering. INCOSE International Symposium, 14(1), pp.1207-1222.

47 Hood, C., 2008. Requirements management. Berlin: Springer.

48 Hood, C. P., 2007. Shinkansen – From Bullet Train to Symbol of Modern Japan (https://books.google.com/books?id=cwbklrvmS-oC). Routledge Contemporary Japan Series. London: Routledge. pp. 18–43. ISBN 978-0-415-32052-8.

49 Hussin, A. A. and Omran, A., 2009. ROLES OF PROFESSIONALS IN THE CONSTRUCTION INDUSTRY, The International Conference on Economics and Administration, Faculty of Administration and Business, University of Bucharest, Romania ICEA – FAA Bucharest, 14-15th November 2009.

50 Hwa, L., Leow, J., Lau, J., Ho, A. and Yong, C., 2020. Design Thinking: The Guidebook. [ebook] Royal Civil Service Commission. Available at: <https://www.rcsc.gov.bt/wp-content/uploads/2017/07/dt-guide-book-master-copy.pdf> [Accessed 6 June 2020].

51 Jamshidi, M., 2009. System of Systems Engineering: Innovation for the 21st Century. Hoboken: Wiley.

52 Kasser, J. E., 2013. *A Framework For Understanding Systems Engineering*. 2nd Edition, The Right Requirement. Cranfield, Bedfordshire, England.

53 Kasser, J. E., 2015. Holistic Thinking – Creating innovative solutions to complex problems. 2nd Edition, The Right Requirement. Cranfield, Bedfordshire, England.

54 Kasser, J. E., 2015b. *Perceptions of Systems Engineering*. Vol. 2, Solution Engineering. CreateSpace Ltd.

55 Kasser, J. E., 2016. *The Nuts And Bolts of Systems*. [video] Available at: <https://www.youtube.com/watch?v=32eejWkyM1E> [Accessed 16 August 2020].

56 Kim, D., 1999. Introduction To Systems Thinking. Pegasus communications, Inc.

57 Koskela, L. & Howell, G. A., 2002. The underlying theory of project management is obsolete. Paper presented at PMI® Research conference 2002: Frontiers of Project Management Research and Applications, Seattle, Washington. Newton Square, PA: Project management Institute.

58 Kossiakoff, A., Sweet, W. N., Seymour, S. J., Biemer, S. M., 2011. Systems Engineering Principles and Practice. 2nd Ed. Canada: Wiley & Sons.

59 Krauss, E., 2014. How Does Operational Readiness Assist in Asset Management: It Surely is an Operations Function? Australian Journal of Multidisciplinary Engineering, 11:1, 1-11.

60 Lessard, D., Sakhrani, V. and Miller, R., 2014. House of Project Complexity - Understanding Complexity in Large Infrastructure Projects. Engineering Project Organization Journal, 4(4), pp.170-192.

61 Leveson, N.G. 2012. *Engineering a Safer World: Systems thinking applied to safety*. Cambridge, MA: MIT Press. Also can download a draft e-copy at: http://mitpress.mit.edu/books/engineering-safer-world.

62 Liedtka, J., 2014. Perspective: Linking Design Thinking With Innovation Outcomes Through Cognitive Bias Reduction. [online] Wiley Online Library. Available at: <https://onlinelibrary.wiley.com/doi/abs/10.1111/jpim.12163> [Accessed 30 May 2020].

63 Lo Storto, C., 2020. Infrastructure megaprojects as complex adaptive systems: a knowledge-based research approach. In: IFKAD 2015 - International Forum on Knowledge Asset Dynamics. [online] Available at: <https://www.researchgate.net/ publication/278019986_Infrastructure_megaprojects_as_complex_adaptive_systems_a_knowledge-based_research_approach/stats> [Accessed 30 May 2020].

64 Love, P.E.D. et al., 2011. What Goes Up, Shouldn't Come down: Learning from Construction and Engineering Failures, The Twelfth East Asia-Pacific Conference on Structural Engineering and Construction, Procedia Engineering 14 (2011) 844-850, 2011.

65 Lukkien, J., 2015. A summary on Systems of Systems. Private Communication, [online] Available at: <https://www.researchgate.net/publication/282734111_A_ summary_on_Systems_of_Systems_private_communication> [Accessed 6 June 2020].

66 Mabelo, P. B., 2016. Application of systems engineering concepts as enhancements to the project life cycle methodology. Masters. University of Witwatersrand.

67 Mabelo, P. B., 2020. How To Manage Project Stakeholders. 1st ed. London: Routledge.

68 Mabelo, P. B., 2020b. Operational Readiness. 1st ed. London: Routledge.

69 Maier, M. and Rechtin, E., 2009. The Art of Systems Architecting. Boca Raton: CRC Press.

70 Maqsood, T.; Finegan, A. D., Walker, D. H. T, 2003. A Soft Approach to Solving Hard Problems in Construction Project Management. In Second International Conference on Construction in the 21st Century. Hong Kong, 10-11 December 2003. Sustainability and Innovation Management and Technology. pp. 313-314.

71 Marnewick, A., Pretorius, L., Pretorius, J., 2014. A South African Perspective of the Requirements Discipline: An Industry Review. South African Institute of Electrical Engineering. Vol. 105(3).

72 Martin, R., 2009. The Design of Business. New York: Harvard Business School Press.

73 Meadows, D. and Wright, D., 2009. Thinking in systems. London: Chelsea Green Publishing.

74 Meadows, D., Randers, J. and Meadows, D., 1972. The limits to growth. White River Junction, VT: Potomac Associates, pp.91-92.

75 Merani, M., 2020. Why Starting Construction Without A Complete Design Is A Bad Idea: Industry Experts. [online] Zawya.com. Available at: <https://www.zawya.com/mena/en/story/Why_starting_construction_without_a_complete_design_is_a_bad_idea_industry_experts-ZAWYA20170702073219/> [Accessed 15 June 2020].

76 Merrow, E., 2011. Industrial megaprojects. 1st ed. New Jersey: Wiley.

77 Mueller-Roterberg, C., 2018. Handbook of Design Thinking: Tips & Tools For How To Design Thinking. Kindle Direct Publishing.

78 NASA. 2011. Beginning Engineering, Science And Technology. [online] Available at: <https://www.nasa.gov/audience/foreducators/best/index.html> [Accessed 15 June 2020].

79 NJDOT, 2016. Design Manual For Bridges And Structures. 6th ed. State of New Jersey Department of Transportation.

80 Oehmen, J., (ed.), 2012. The guide to lean enablers for managing engineering programs, version 1.0. Cambridge, MA: Joint MIT-PMI-INCOSE Community of Practice on Lean in Program Management. [online] Available at: http://hdl. handle.net/1721.1/70495 [Accessed 24 Oct 2019].

81 Okoh, P., Schjølberg, P., Wilson, A., 2016. Review and Application of ISO 5500 Asset Management Standards in relation to Maintenance management, Norsk Forening for Vedlikehold (NFV).

82 Panther Movie Productions, 2014. Megafactories - Pagani Huayra. [video] Available at: <https://www.youtube.com/watch?v=tb1Q0A4J1jc&list=PLlwk_3JcQrSEcNEEW No8ZT4cc-bxwEVEG&index=21&app=desktop> [Accessed 4 July 2020].

83 Pararas-Carayannis, C., 2020. Rion-Antirion Bridge: An Engineering Marvel Spanning The Gulf of Corinth, Greece. By Carolyn Pararas-Carayannis. [online] Info-now.org. Available at: <http://info-now.org/Greece/Rion-AntirionBridge.php> [Accessed 15 June 2020].

84 Petroski, H., 1992. To Engineer Is Human. New York: Vintage Books.

85 Popova, M., 2020. French Polymath Henri Poincaré On How The Inventor's Mind Works, 1908. [online] Brain Pickings. Available at: <https://www.brainpickings.org/2012/06/11/henri-poincare-on-invention/> [Accessed 15 June 2020].

86 Project Management Institute, 2017. Agile Practice Guide. Newton Square, Pennsylvania: Project Management Institute.

87 Project Management Institute. 2013. A Guide to the Project Management Body of Knowledge (PMBoK Guide). Project Management Institute, 5th Ed.

88 ProRail, Rijkswaterstaat, Bouwend-Nederland, ONRI, 2013. Guideline Systems Engineering for Public Works and Water Management, Version 3.

89 ProRail, 2013b. Guidelines for Systems Engineering Within the Civil Engineering Sector. Version 3. Available at http://www.leidraadse.nl/assets/files/downloads/LeidraadSE/V3_EN/boek_Leidraad_SE_ENG_3_72dpi-def.pdf. [Accessed 25 June 2016].

90 Robertson, S. and Robertson, J., 2005. *Requirements-Led Project Management*. Boston: Addison-Wesley.

91 Roy, R., 1993. Case studies of creativity in innovative product development. Design Studies, [online] 14(4), pp.423-443. Available at: <https://www.researchgate.net/publication/239362023_Case_studies_of_creativity_in_innovative_product_development> [Accessed 15 June 2020].

92 Rumane, A. R., 2018. Quality Management in Construction Projects, Second Edition, CRC Press, Boca Raton, Florida.

93 Russia Today (RT), 2014. "Berlin airport hired fake engineer on key position". [Online Video]. 29 June 2014. Available from:

<https://www.youtube.com/watch?v=
M0_DPXgDJ10> [Accessed 14 September 2020].

94 Ryen, E., 2008. Overview of the systems engineering process. Intelligent Transportation Systems, North Dakota, Department of Transportation. [online] Available at: https://ww w.dot.nd.gov/divisions/maintenance/docs/ OverviewOfSEA.pdf [Accessed 25 June 2016].

95 Ryschkewitsch, M., Schaible, D. and Larson, W., 2009. THE ART AND SCIENCE OF SYSTEMS ENGINEERING. Systems Research Forum, [online] 03(02), pp.81-100. Available at: <https://www.worldscientific.com/doi/10.1142/S1793966609000080> [Accessed 15 June 2020].

96 San Cristobal, J., Carral, L., Diaz, E., Fraguela, J. and Iglesias, G., 2018. Complexity and Project Management: A General Overview. Complexity, 2018, pp.1-10.

97 Sandia Laboratories., 2011. Complex Adaptive System of Systems (CASoS) Engineering Framework Version 1.0. Sandia National Laboratories Report SAND 2011-8793.

98 Schuetze, C., 2020. Berlin's Newest Airport Prepares for Grand Opening. Again. The New York Times, [online] Available at: <https://www.nytimes.com/2020/04/29/ world/europe/berlin-brandenburg-airport-opening.html> [Accessed 16 August 2020].

99 Scott, Z., 2012. 9 Laws of Effective Systems Engineering. White Paper. Vitech Corporation. Available at: http://www.vitechcorp.com/resources/white_papers/ 9_Laws_of_Effective_Systems_Engineering.pdf. [Accessed 25 June 2016].

100 Scott, Z., 2014. "Engineers Behaving Badly". [Online Video]. 26 February 2014. Available from: https://www.youtube.com/watch?v=_x-zTOiPXTU. [Accessed 29 June 2016].

101 Shrady, N., 2003. Tilt: a skewed history of the Tower of Pisa.

102 Shrivastava, P., 2015. House of Quality: An Effective Approach to Achieve Customer Satisfaction & Business Growth in Industries. International Journal of Science and Research (IJSR), [online] 6(14), pp.1365 - 1371. Available at:

<https://www.ijsr.net/archive/v5i9/ART20161787.pdf> [Accessed 6 June 2020].

103 Simon, H. A., 1956. Rational Choice and the Structure of the Environment, Psychological Review Vol. 63, No 2, 1956, Carnegie Institute of Technology.

104 Sivaloganathan, S. and Al-Marzouqi, A. H., 2014. Innovative Teaching of Product Design and Development in an Engineering Management Program, 121st ASEE Annual Conference and Exposition, Indianapolis, Indiana, June 15-18 (2014), American Society for Engineering Education, 2014.

105 Slack, N., Brandon-Jones, A. and Johnston, R. (2010). *Operations management*. 6th Ed. Prentice Hall-Financial Times.

106 Slocum, A., 2008. Fundamentals of Design Topic 1 Design Is A Passionate Process. 1st ed. [ebook] Available at: <https://web.mit.edu/2.75/fundamentals/FUNdaMENTALs%20Book%20pdf/FUNdaMENTALs%20Topic%201.PDF> [Accessed 15 June 2020].

107 Snee, R. and Rodebaugh, W., 2020. Failure Modes And Effects Analysis (FMEA) Tool | IHI - Institute For Healthcare Improvement. [online] Ihi.org. Available at: <http://www.ihi.org/resources/Pages/Tools/FailureModesandEffectsAnalysisTool.aspx#:~:text=Failure%20Modes%20and%20Effects%20Analysis%20(FMEA)%20is%20a%20systematic%2C,most%20in%20need%20of%20change.> [Accessed 15 June 2020].

108 Standards and Specifications Specialist, 2018. Technical Design Requirements For Alberta Infrastructure Facilities. 4th ed. [PDF] Alberta Infrastructure Technical Services Branch. Available at: <http://www.infrastructure.alberta.ca/Content/doc Type486/Production/TechDesignRequirements.pdf> [Accessed 15 June 2020].

109 Stanfordonline, 2013. Design Thinking: Training Yourself To Be More Creative. [video] Available at: <https://www.youtube.com/watch?v=34EuT2KH2Lw> [Accessed 16 August 2020].

110 Stephan, D., 2013. Statistics For Managers Using Microsoft Excel. Upper Saddle River: Pearson Education.

111 Stille, Alexander. “The World’s Oldest Papyrus and What It Can Tell Us About the Great Pyramids” (http://www.smithsonianmag.com/history/ancient-egypt-shipping-mining-farming-econo my-pyramids-180956619). Archived <https://web.archive.org/web/20150928173821/http://www.smithsonianmag.com/history/ancient-egypt-shipping-mining-farming-economy-pyramids-180956619/> from the original on 28 September 2015. Retrieved 16 August 2020.

112 Sullivan, L., 1896. The tall office building artistically considered. Lippincott’s Magazine, [online] Available at: <https://ocw.mit.edu/courses/architecture/4-205-analysis-of-contemporary-architecture-fall-2009/readings/MIT4_205F09_Sullivan.pdf> [Accessed 15 June 2020].

113 Systems engineering handbook, 2007. 1st ed. [PDF] Washington, DC: NASA.
[online] Available at:
https://www.nasa.gov/sites/default/files/atoms/files./
nasa_systems_engineering_handbook.pdf [Accessed 29 April 2018].

114 Tapke, Jennifer & Muller, Allyson & Johnson, Greg & Sieck, 2001. Josh, House of Quality: Steps in Understanding the House of Quality.

115 The Washington Post, 1982. A Tale of Two Projects: One Winner, One Loser. [online] Available at: <https://www.washingtonpost.com/archive/opinions/1982/04/25/a-tale-of-two-projects-one-winner-one-loser/fc3f76b7-e608-47d1-95aa-974ae68176f8/> [Accessed 15 June 2020].

116 Tischler, L. (2009). Ideo's David Kelley on Design Thinking. Fast Company, 132. [online] Available at: <https://www.fastcompany.com/1139331/ideos-david-kelley-design-thinking>. [Accessed 15 June 2020]

117 Van Ruijven, L., 2013. Ontology for Systems Engineering. Procedia Computer Science, [online] 16, pp.383–392. Available at: <https://www.researchgate.net/publication/257719545_Ontology_for_Systems_Engineering> [Accessed 15 June 2020].

118 Vermaas, P., 2010. Philosophy And Design. Dordrecht: Springer.

119 Wallace, K., Clarkson, J. and Huhtala, M., 2005. Engineering Design. Cambridge: Engineering Design Centre, University of Cambridge.

120 Whelton, M. and Ballard, G., 2002. Wicked problems in project definition. Proceedings of the 10th Annual Conference International Group for Lean Construction, Brazil. Available at: http: //www.lean construction.org/media/docs/

WickedProblemsinProjectDefinitionIGLC10.pdf [Accessed 25 June 2016].

121 Youker, R., 1998. Defining the hierarchy of project objectives. In: IPMA Conference. [online] Available at: http: //www.ipma-usa. org/articles/m_hierobjs.pdf [Accessed 18 October 2019].

122 Zimmer, T. W., Scarborough, N.M. and Wilson, D., 2008. "Essentials of Entrepreneurship and Small Business Management," 5th Edition, Prentice-Hall, Upper Saddle River.

Additional Readings

[a] Bazaati, S. and Beheshtizadeh, D., 2020. Using Quality Function Deployment (QFD) in the Construction Company. Journal of Applied Environmental and Biological Sciences, [online] 7(6), pp.7-16. Available at: <https://www.textroad.com/pdf/JAEBS/J.%20Appl.%20Environ.%20Biol.%20Sci.,%207(6)7-16,%202017.pdf> [Accessed 4 July 2020].

[b] Burnham, R., 2014. An Overview of Complexity Theory For Project Management. [PDF] Notre Dame de Namur. Available at: <https://www.academia.edu/6908046/An_Overview_of_Complexity_Theory_for_Project_Management> [Accessed 30 May 2020].

[c] DOD. *MIL-STD-499A (USAF), Engineering Management.* 1974 (or more recent).
[Superseding the original version of *MIL-STD-499 (USAF) of the 17th July 1969].*

[d] KPMG International, 2016. Assessing The True Value of Infrastructure Investment. 1st ed. [ebook] Switzerland: KPMG International Cooperative. Available at: <https://home.kpmg/content/dam/kpmg/ge/pdf/2016_02/

True%20Value%20of%20InfraInvestment.pdf> [Accessed 4 July 2020].

[e] Lucerne School of Art and Design, 2016. The Role of Design Management. [PDF] Lucrene. Available at:

<https://cdn.ymaws.com/www.dmi.org/resource/resmgr/pdf_files/Lucerne_School_of_Art_&_Desi.pdf> [Accessed 4 July 2020].

[f] Pilsbury, L. A. and Reese, L. A., 2001. Recognizing project management excellence – making it crystal clear. Paper presented at Project Management Institute Annual Seminars and Symposium, Nashville, TN. Newtown Square, PA: Project Management Institute.

[g] Slade D. and Davies N., 2017. How to design an infrastructure strategy for the UK. Institute for Government. Available at www.instituteforgovernment.org.uk/publications/how-design-infrastructurestrategy-uk [Accessed 29 November 2017].

INDEX

C

D

L

M

N

O

S

T

U

V